Wavelength Division
Multiplexing

Prentice Hall International Series in Optoelectronics

Consultant editors : John Midwinter, University College London, UK
Bernard Weiss, University of Surrey, UK

Fundamentals of Optical Fiber Communications
W. van Etten and J. van der Plaats

Optical Communication Systems
J. Gowar

Optical Fiber Communications : Principles and Practice (Second Edition)
J.M. Senior

Lasers : Principles and Applications
J. Wilson and J.F.B. Hawkes

Optoelectronics : An Introduction (Second Edition)
J. Wilson and J.F.B. Hawkes

Wavelength Division Multiplexing

J.P. LAUDE

Preface by André MARÉCHAL

Published with the support of Ministère de la Recherche et de l'Espace (DIST), France.

Prentice Hall

New York   London   Toronto   Sydney   Tokyo   Singapore

MASSON

Paris   Milan   Barcelone

This edition published 1993 by
Prentice Hall International (UK) Limited
Campus 400, Maylands Avenue
Hemel Hempstead
Hertfordshire, HP2 7EZ
A division of
Simon & Schuster International Group

© Masson, Paris, 1993

This book was originally published under the title :
Le multiplexage de longueurs d'ondes by Jean-Pierre Laude
in the series "Lasers et optoélectronique", © 1992, Masson, Paris.

All rights reserved. No part of this publication may be reproduced, stored in a retrieval system, or transmitted, in any form, or by any means, electronic, mechanical, photocopying, recording or otherwise, without prior permission, in writing, from the publisher.
For permission within the United States of America
contact Prentice Hall Inc., Englewood Cliffs, NJ 07632.

Printed and bound in France by Normandie Impression.

Library of Congress Cataloging-in-Publication Data

Available from the publisher

British Library Cataloging in Publication Data

A catalogue record for this book is available from the British Library

ISBN 0-13-489865-6 (hbk)

Contents

Foreword		ix
Preface		xi
Symbols		xii
Introduction		1

Chapter 1   **Fibre propagation: theoretical background using electromagnetic formalism**   3

     1.1   Introduction   3
     1.2   Propagation modes in step index fibres   4
     1.3   Useful approximations for single-mode fibres with graded index core   9
     1.4   Remarks on dispersion properties   11
     1.5   Depressed Inner Cladding (DIC) fibres and other modified core fibres   12
     1.6   Soliton propagation   13

Chapter 2   **Multimode fibres: geometrical optics analysis**   15

Chapter 3   **Wavelength division multiplexing: basic principles**   19

Chapter 4   **Multi/demultiplexers: main characteristics**   25

Chapter 5   **Multidielectric filters used in multiplexing**   27

Chapter 6   **Multidielectric filter devices**   31

Chapter 7   **Gratings for multiplexing**   37

     7.1   Introduction   37
     7.2   Efficiency versus wavelength   39

         7.2.1   Plane reflection gratings   39
         7.2.2   Transmission gratings   45
         7.2.3   Concave gratings   46
         7.2.4   Practical efficiency   47

7.3	Bandwidth of grating devices	48
	7.3.1 Devices with input and output single-mode fibres	48
	7.3.2 Devices with a single-mode entrance fibre and an exit slit or a diode array with rectangular pixels	53
	7.3.3 Devices with a single-mode entrance fibre and multimode step index exit fibres	57
	7.3.4 Devices with a single-mode entrance fibre and multimode graded index exit fibres	58
	7.3.5 Coupling in a device with small aberrations	59

Chapter 8 Grating microoptic devices 60

8.1	Grating multiplexers using all multimode fibres or grating demultiplexers with a single-mode entrance fibre and multimode exit fibres	60
8.2	All single-mode fibre grating multi/demultiplexers	64
8.3	Thermal drift	67

Chapter 9 Multiplexers using wavelength selective coupling between fibres 68

Chapter 10 Trend towards 'integration' of the wavelength division multiplexers and demultiplexers 73

10.1	Introduction	73
10.2	Source integration in multiplexing	74
10.3	Integration of detectors in demultiplexing	77
10.4	Multi/demultiplexing on planar optical waveguides	78
	10.4.1 Devices using lithium niobate (or lithium tantalate)	78
	10.4.2 Devices using amorphous dielectric or semi-conductor waveguides	79

Chapter 11 Some techniques used in wavelength division multiplexing 92

11.1	Light Emitting Diode (LED) spectral slicing	92
	11.1.1 Introduction	92
	11.1.2 Theoretical analysis	94
	11.1.3 Calculation of the spectral filtering losses of real systems	100
11.2	Shared optical function technology wavelength division multiplexing and coupling ('SOFT WDM' and 'SOFT couplers')	105
	11.2.1 General principle	105
	11.2.2 The particular case of grating multiple multiplexers	106
	11.2.3 Remark	110

	11.3	Tuneable demultiplexers	111
		11.3.1 Fabry-Perot devices	111
		11.3.2 Tuneable acoustooptic filters	113
	11.4	Wavelength multiplexers using polarizing beam splitters	114
	11.5	Neural network multiplexers	116
	11.6	Photodiode array demultiplexers	117
	11.7	Short wavelength transmission over single-mode fibres optimized for long wavelengths	118
Chapter 12		**Wavelength division multiplexing and optical amplification**	**119**
	12.1	Introduction	119
	12.2	Tuneable Fabry-Perot amplifiers	119
	12.3	Travelling wave semi-conductor amplifiers	120
	12.4	Brillouin scattering amplifiers	120
	12.5	Raman scattering amplifiers	121
	12.6	Rare earth doped fibre optic amplifiers	121
Chapter 13		**Application of wavelength division multiplexing to telecommunication networks**	**126**
	13.1	Introduction	126
	13.2	Physical and virtual topologies	130
	13.3	Passive optical networks	131
		13.3.1 A first generation of applications	131
		13.3.2 An evolution towards single-mode fibre networks and towards more wavelengths	132
	13.4	Evolution towards coherent optical networks	139
		13.4.1 Coherent detection principle	139
		13.4.2 Multichannel coherent detection	139
		13.4.3 Detectivity in coherent detection	140
		13.4.4 Coherent subcarrier multiplexing (SCM)	140
		13.4.5 Coherent optical multiplexing in FSK heterodyne detection	141
		13.4.6 Homodyne and heterodyne PSK coherent optical multiplexing	143
		13.4.7 Source wavelength stabilization and wavelength standards	144
	13.5	Wavelength division multiplexing for optical switching and routeing	144
		13.5.1 Introduction	144
		13.5.2 'Broadcast and select' networks	145
		13.5.3 Wavelength routeing networks	146
		13.5.4 Remarks	148

viii   Contents

                13.6   Miscellaneous point-to-point transmission links   148

                        13.6.1 Full duplex links   149
                        13.6.2 Links combining frequency multiplexing
                              and/or digital techniques and
                              wavelength division multiplexing   149
                        13.6.3 Evolution of wavelength division multiplexing
                              for trunk lines   151
                        13.6.4 Miscellaneous remarks   151

Chapter 14   **Other applications of wavelength division multiplexing**   **152**

                14.1   LED linewidth narrowing for reduction of
                            chromatic dispersion   152
                14.2   Transmission protection   153
                14.3   Optical interconnection   153
                14.4   Wavelength division multiplexers for optical pump coupling
                            and amplified signal filtering in doped fibre lasers   153
                14.5   Dispersion measurements   154
                14.6   Sensors   154
                14.7   Industrial control and sensor networks   155
                14.8   Telespectroscopy   157
                14.9   Multiplexing of radar signals   161
                14.10 All optical image transmission   162

Chapter 15   **Some limitations of wavelength division multiplexing**   **163**

                15.1   Crosstalk effects   163
                15.2   Polarization effects in multi/demultiplexers   164
                15.3   Crosstalk due to Raman conversion   164
                15.4   Crosstalk due to other non-linear effects   165
                15.5   Minimum channel spacing related to 'uncertainty'
                            relationships   165
                15.6   Effect of spectral filtering on multimode lasers   167
                15.7   Directly modulated DFB LD spectral spread   168
                15.8   Wavelength division multiplexing of solitons   169
                15.9   Some other remarks   170

Conclusion   **172**

Bibliography   **174**

Index   **208**

> *Peu de tout. Puisqu'on ne peut être universel*
> *en sachant tout ce qui se peut savoir sur tout,*
> *il faut savoir peu de tout...*[1]
>
> Blaise Pascal

Foreword

Wavelength division multiplexing (WDM) is past the basic research stage and requires the attention of the optical fibre communication engineering field for its three main advantages: it increases fibre capacity without incurring extra installation costs, it allows for flexible expansion of systems and it provides an evolutionary path for future services on an existing network.

This book is the fruit of more than ten years of study in the field and was first written for the students of the Ecole Nationale Supérieure des Télécommunications in Paris in 1989. I have tried to provide the necessary theory, background and practical information for the intended readers, who may be scientists, designers, development engineers, technicians or graduate students. It is expected that the book will be useful in the design of optical networks as well as being one of the bases for a Master of Science course on optical telecommunications.

While writing, I have sought to keep in mind at least three ideas. The first one is to set down the definitions and bases of optical multiplexing. The second one is to summarize the recent developments, thus a large proportion of the book is a review of other books. (I have tried to give exhaustive references throughout the text. To any contributors to the WDM field, not finding their names in these references, I humbly apologize; it was not possible to be completely exhaustive). Finally, I have tried to give some characteristics of networks using it.

---

[1] A little of everything. As it is not possible to be universal by knowing all that can be known on everything, we must know a little of everything...

I would like to record my thanks to all those who contributed to the preparation of this book, in particular Mrs Claude Lemaire, my secretary, for her tremendous job on the manuscript, Vincent Laude, my son, and Pierre Augais, my brother-in-law, for helpful cooperation, the scientists and the management of Jobin-Yvon, the RACE 1036 WTDM research group and personnel from the Centre National d'Etudes des Télécommunications, who collaborated with us in the field. I also wish to thank Professor Patrick Bouchareine, who read the first manuscript, and John Zubrzycki and Vernon Fraval who improved its English version.

Christine, my wife, and our children, Jean-Christophe, Vincent, Thomas, Blandine, Marjolaine and Grégoire have been a great source of support and encouragement. They found life dull during my long week-ends of manuscript composition. I am grateful to them for their love and patience.

Jean-Pierre Laude [2]
Docteur Ingénieur
Université Paris XI
Directeur Scientifique
de ISA Jobin-Yvon

---

[2] J.-P. Laude, Les Graveriots, 91690 St Cyr la Rivière, France.

Preface

Telecommunications make wide use of optical techniques where the carrier wave belongs to the classical optical domain. The wave modulation allows signal transmission and its very high frequency permits high information bit rates in telephony as well as in television. In fact, the bit rate can be increased further, using several carrier waves that are propagating without significant interaction on the same cable if the frequencies corresponding to the different carriers are adequately spaced. Such is the wavelength division multiplexing principle. It must be understood that the overall system design needs an optimization of each component characteristic and a good implementation compatibility in this new technique. Thus, optical amplification on a doped fibre (for instance, using erbium) is now possible. With this method, one can avoid transforming the optical signal into an electrical signal, and conversely in the amplifier stage. It really seems that optical amplification will soon be practically compatible with wavelength division multiplexing.

Jean-Pierre Laude's book is a basic text, well informed and written by someone who has contributed fully through his research work and his teaching to the progress of modern optoelectronic systems, not only in the telecommunications area but also in many other high-technology sectors. Among these sectors, let us mention process control in complex manufacturing where optical fibres are not only used for information transmission but also as sensors of various physical parameters. The fact that wavelength division multiplexing will be widely and universally used is unquestionable.

André Maréchal
Member of the Académie des Sciences
Emeritus professor at Paris Sud University
Former Délégué général
à la Recherche Scientifique

Symbols

a	core radius
a'	inner cladding radius in RIC and DIC fibres
$A(x'y')$	amplitude function in exit plane
a_e	equivalent core radius
$b(V)$	normalized propagation constant
B_{IF}	intermediate frequency filter bandpass
c	speed of light
$c(h)$, C_0	coupling coefficients
C	carrier power
C/N	carrier to noise ratio
d	distance between grooves
D_{ij}	optical crosstalk of channel i on channel j in dB
e	thickness of a layer or a slit (Chapter 7)
E_n, E_c, E_A	energy
E_z, E_y, E_o	electric fields
$e^{-i\omega t}$	exp $-i\omega t$

Symbols

F	optical flux
f(x')	function f of the variable x'
F	multiplexer spectral transmission function
g	parameter
g(x')	function g of the variable x'
$\tilde{g}(u)$	Fourier transform of g(x')
G	emitter intensity function
G_a	gain
h	Planck constant
h_0	groove depth of a grating
H	high index (Chapters 5 and 6)
H_x	magnetic field along x
i, i'	incident, diffraction angles from M (Figure 27)
I	optical intensity
J_0, J_1, J_2	Bessel functions of orders 0, 1 and 2
k	in Chapter 7, integers: −2, −1, 0, 1, 2
k, k_a, k_o	wave vector module, optical in general, acoustic or optical extraordinary ($k = 2\pi/\lambda_0$)
K_0, K_1, K_2	modified Bessel functions of order 0, 1 and 2
L, L_0	lengths
n	refractive index (however, n is an integer in Section 11.2.2)
n_1	core index

n_2	cladding index
n_3	external index in DIC fibres
n_s	shot noise
n_c	carrier noise
n_t	thermal noise
N_0	total number of grooves on a grating
N	normalized frequency
p	number of fibres on a 'SOFT WDM'
P_j	loss in channel j (in dB)
r	radius
r, φ	polar coordinates in a fibres section
R_1, R_2	integration radius parameters
R_s	resolution
R_ω	relative spectral width
R_{F1}, R_{F2}	reflectivity
S	substrate
$S(y)$	modulus of acoustic wave signal along y direction
$\hat{S}(y,t)$	acoustic wave signal
S_1, S_2	integration surfaces
t	Fabry-Perot thickness
T	transmission coefficient
$T(x'y')$	optical transmission in exit plane

Symbols

TF, TF−1	Fourier transform and reverse transform
u, υ	transverse mode propagation parameters
w	mode radius in general
x y z x′ y′ z′	scalar coordinates in object and image space respectively (z, z′ along propagation, except in Section 11.3.2)
Z_0	vacuum impedance
Z_{s0}	soliton period
$\emptyset$	fibre core diameter
α, α_i, α', α_j	incident, diffracted angles from N (Figures 27 and 86)
β	propagation constant
$\gamma = 25°$	blaze angle (except in Chapter 1: parameter)
Δ	small difference
Δ_0	optical path difference
$\Delta\lambda$, $\Delta\omega$, $\Delta\zeta$, $\Delta\nu$	small variation of wavelength, pulsation, time, optical frequency, respectively
ζ	group delay per unit length
η	quantum efficiency
θ	ray acceptance cone half-angle, outside the fibre
λ	wavelength
λ_c	cutoff wavelength
ν	optical frequency $\lambda = \dfrac{c}{\nu}$

$\sigma = \dfrac{1}{\lambda}$	wavenumber
τ	pulse width
υ	transverse mode propagation parameters
φ, r	polar coordinates in a fibre section
Φ	optical power
ϕ	angle $\vec{r}\ O\vec{x}$ in Chapter 2
Ψ	normalized electric field
$\omega, \omega_a, \omega_{ex}, \omega_{or}$	propagation angular velocities, respectively: in general, acoustic, extraordinary, ordinary
ω_0, ω'_0	mode radius in object plane, in image plane

> *Tous les rêves dynamiques, des plus violents aux plus insidieux, du sillon métallique aux traits les plus fins, vivent dans la main humaine, synthèse de la force et de l'adresse ...*[1]
>
> Gaston Bachelard

Ruling machine making a 600 lines per millimetre diffraction grating. This metallic master will be used to make copies for wavelength division multiplexers.

---

[1] **Every dynamic dream, from the most violent to the most insidious; from the metallic grooves to the finest lines, lives in the human hand as a synthesis of strength and dexterity.**

*Bien entendu, la recherche et la découverte
ne vont pas sans un certain désordre*[1]

Hermann Kahn

Introduction

The optical multiplexing concept is now quite old. It dates back to at least 1958 [1] [2]. But it was necessary to wait until 1977 to obtain the first practical solutions put forward by Tomlinson and Aumiller [3]. Wavelength division multiplexing, also called optical frequency multiplexing (we will see when later), is perhaps a rather unfamiliar domain for some people. However, it has been profitable to almost all of us, a fact which has probably gone unnoticed. Let us choose an example : the retransmission of the 1992 Winter Olympic Games [4]. Several links between the competition sites and Albertville were optical and wavelength division multiplexed (using the Thomson CSF links of [350] and Jobin Yvon components (Figure 112)). Still more than that anecdotal example, generally speaking, over the past decade, we have witnessed a rapid growth in optical telecommunications. At first, it was used for long distance links. For shorter distances, important applications were found in videobroadcasting. Moreover, high-definition television bandwidth requirements (in Gb/s) have renewed interest in optical fibres in such transmissions. Nowadays, optical transmission is finding growing applications in short and medium-distance professional links as well as in local networks, and seems capable of taking a prominent part in telephony customer networks. The links between supercomputers with high-resolution graphic terminals (800 Mb/s) will probably use optical fibres (they are already in use in a few modern networks).

The bit-rate to be transmitted becomes higher and higher and, despite the fact that single-mode fibre has a substantial bandpass, the wavelength division multiplexing that corresponds to the superimposition of optical signals at different wavelengths on a single fibre, often becomes, as a matter of fact, the best choice for a capacity increase at lower cost. Furthermore, it also has many other advantages, for example in routeing and switching. Lastly, its applications extend further and further beyond the optical telecommunication domain.

---

[1] That's settled, research and discovery won't do without some disorder.

We did our best to organize the following text into a logical sequence of chapters. But our work encompasses a field in which H. Kahn's remark, given at the beginning of this introduction, is particularly relevant. So, it is probably just as well that we, too, were unable to avoid some disorder traces. It is possible to read this book in different ways according to the reader's main interest: whether the priority is theoretical aspects, components or applications. The first category of readers may skip over the main part of Chapters 6 and 8. The two other categories of readers may begin at Chapter 3. Most components are described in Chapters 5 to 10 and most applications in Chapters 13 and 14. However, from our point of view, such an arbitrary partition is not desirable and we are convinced that reading chapters sequentially would be more profitable.

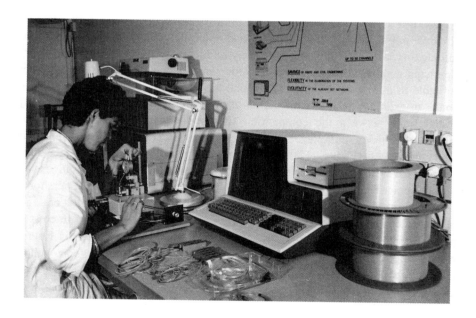

Wavelength division multiplexer tests.

History and exact science he must learn by laborious reading.

Ralph Waldo Emerson

CHAPTER 1

Fibre propagation: theoretical background using electromagnetic formalism

1.1 Introduction

The light-guiding structure in an optical fibre is generally small: the fibre core diameter is only a few tens of wavelengths, even a few wavelengths. So it is necessary to use the Maxwell equations to define the electromagnetic field propagation conditions in the fibre rigorously. It has been demonstrated that the electromagnetic field can be decomposed into propagation modes, the meaning of which will be given hereafter. If the wavelength can only support one propagation mode, the fibre is called single-mode. Such a fibre has a typical diameter of 8 to 12 µm. Single-mode fibre is preferred to multimode fibre in telecommunication networks due to several advantages: higher bandwidth, lower intrinsic losses, compatibility with integrated optics and lower price (except in a few examples such as plastic fibre). In this book, we will concentrate on the analysis of single-mode fibre multiplexers. However, the case of multimode fibre multiplexers will be set out. Maxwell's theory is indispensable for the description of propagation in single-mode fibres, but, in multimode fibres, geometrical approximations can be used, otherwise the modal analysis becomes inextricable. Thus, for a fibre with an index difference between core and cladding of 10^{-2} and with a core diameter of 200 µm, about 3000 modes can be counted. We used documents [5] to [11] as our main theoretical background sources for this chapter. A much more detailed analysis can be found in these papers.

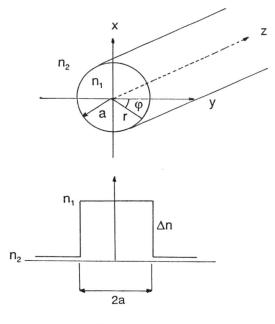

Figure 1
Step index fibre.

1.2 Propagation modes in step index fibres

In such a fibre, the core index n_1 is constant and larger than the cladding index n_2 (Figure 1) over the core radius 2a. Analysis of Maxwell's equations for lossless propagation gives a set of solutions called propagation modes. The electric field corresponds to solutions such as $E(r, \varphi)\exp i(\omega t - \beta z)$, where ω is the angular velocity and β the propagation constant, r and φ are given in Figure 1. Generally, the modes are hybrid and must satisfy the limit conditions for six components of the electric and magnetic fields. The simplest case occurs when there is no variation of the fields with φ. In such a simple case, TE and TM solutions[2] are found; only two coordinates (x,z or y,z) are sufficient to describe the fields as in the planar waveguide case. The number of modes that can exist is a function of the fibre core radius to wavelength ratio and of the index difference $\Delta n = n_1 - n_2$.

There is a finite number of propagating guided modes corresponding to discrete values β_g (g for guided). These values are such that $n_2 k < |\beta_g| < n_1 k$ where n_2 is the external index, n_1 the core index and $k = 2\pi/\lambda_0$ (λ_0 being the wavelength).

---

[2] TE: transverse electric; TM: transverse magnetic.

Applying Maxwell's equations to the structure gives solutions relatively simply expressed as a function of $n_1^2 k^2 - \beta_g^2$ inside the guide ($0 \leq r < a$) and $n_2^2 k^2 - \beta_g^2$ outside the guide ($r \geq a$). We write

$$n_1^2 k^2 - \beta_g^2 = \frac{u_g^2}{a^2} \quad \text{and} \quad n_2^2 k^2 - \beta_g^2 = -\frac{v_g^2}{a^2}$$

u_g/a is a transverse constant corresponding to the propagation vector in the core, v_g/a corresponds to the transverse exponential decay in the cladding.

If a normalized frequency V is defined by the following formula:

$$V = ak \sqrt{n_1^2 - n_2^2} \quad (\text{with } k = 2\pi/\lambda_0)$$

the fields can be expressed as a function of the normalized frequency V.

Considering the HE_{11} mode:

It can be demonstrated that u and v satisfy the two following equations simultaneously:

$$u^2 + v^2 = V^2 \quad \text{and} \quad u \frac{J_1(u)}{J_0(u)} = v \frac{K_1(v)}{K_0(v)}$$

where J_0 and J_1 are the Bessel functions of orders 0 and 1 and K_0 and K_1 are the modified Bessel functions of orders 0 and 1. The electric field is given by:

$$E_{y,x} = E_0 \begin{cases} \dfrac{J_0(u\, r/a)}{J_0(u)} & 0 \leq r \leq a \\ \dfrac{K_0(v\, r/a)}{K_0(v)} & r \geq a \end{cases}$$

$$E_z = -i \frac{E_0}{ka\, n_2} (\sin \varphi, \cos \varphi) \begin{cases} \dfrac{u\, J_1(u\, r/a)}{J_0(u)} & 0 \leq r \leq a \\ \dfrac{v\, K_1(v\, r/a)}{K_0(v)} & r \geq a \end{cases}$$

$$Hz = -i \frac{E_0}{ka\, Z_0} (\cos \varphi, \sin \varphi) \begin{cases} \dfrac{u\, J_1\, (u\, r/a)}{J_0\, (u)} & 0 \leq r \leq a \\ \dfrac{\upsilon\, K_1\, (\upsilon r/a)}{K_0\, (\upsilon)} & r \geq a \end{cases}$$

Considering a perfectly circular geometry the two HE_{11} modes along Ox or Oy are degenerated in the so-called LP_{01} (LP for linear polarization).

Note

Taking into account the normalization conditions of the power, with Z_0 being the vacuum impedance, this corresponds to:

$$E_0 = \frac{u}{V} \frac{K_0\, (\upsilon)}{K_1\, (\upsilon)} \left(\frac{2\, Z_0}{\pi a^2 n_2} \right)^{1/2}$$

Considering the second-order modes (TM_{01}, TE_{01}, HE_{21}):

u_1 and υ_1 correspond to solutions that satisfy the two following equations simultaneously:

$$u_1^2 + \upsilon_1^2 = V^2 \quad \text{and} \quad u_1 \frac{J_2\, (u_1)}{J_1\, (u_1)} = \frac{K_2\, (\upsilon_1)}{K_1\, (\upsilon_1)}$$

where J_2 and K_2 are the modified Bessel functions of order 2. The electric field can be written as:

$$Ey, x = E_1 (\cos \varphi, \sin \varphi) \begin{cases} \dfrac{J_1\, (u_1\, r/a)}{J_1\, (u_1)} & 0 \leq r \leq a \\ \dfrac{K_1\, (\upsilon_1\, r/a)}{K_1\, (\upsilon_1)} & r \geq a \end{cases}$$

(The modes are also called LP_{11}). The longitudinal components are weak and there are four possible combinations for (x, y) and (cos φ, sin φ).

The calculated values of u, u_1, u_2 ... (and consequently υ, υ_1, υ_2 ...) against normalized frequency are given in Figure 2 for different modes. The values of the wavelength, the core radius and the indices being known, will enable V to be found and hence deduction of the value of u_n from the diagram and therefore of υ_n for the different modes.

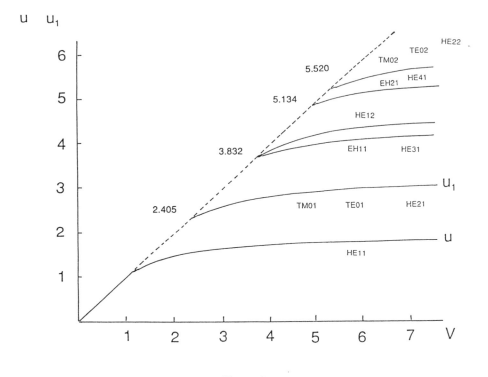

Figure 2
Transverse propagation constants of the modes as a function of normalized frequency.

As V increases, other modes appear. The cutoff frequency of a mode corresponds to the intersection with the dashed line. The cutoff frequency of modes of second order is V = 2.405. This corresponds to a cutoff wavelength λ_c above which only the degenerated mode HE_{11} propagates without losses.

$$\lambda_c = \frac{2\pi a \sqrt{n_1^2 - n_2^2}}{2.405}$$

The cutoff frequency of the degenerated mode HE_{11} being zero, we obtain the so-called single-mode propagation for $\lambda_c < \lambda < \infty$

Field distribution Gaussian approximation

It can be demonstrated that the field profile of the LP_{01} mode can be represented by a Gaussian profile with good approximation:

$$Hx = \frac{2}{w}\left(\frac{n_2}{Z_0 \pi}\right)^{1/2} \exp\left[-\left(\frac{r}{w}\right)^2\right]$$

where Z_0 is the vacuum impedance, n_2 is the core index and n is the transverse distance from the core axis; w must be adjusted.

For $0.8 \leq \lambda/\lambda_c \leq 2$, we obtain a good approximation, with accuracy better than 1% on launching power efficiency, if we take for w the value given by the following equation:

$$\frac{w}{a} = 0.65 + 0.434\left(\frac{\lambda}{\lambda_c}\right)^{3/2} + 0.0149\left(\frac{\lambda}{\lambda_c}\right)^6$$

We also have:

$$Ey = \frac{Z_0}{n_2} Hx \quad \text{and} \quad Ey = \frac{2}{w}\left(\frac{Z_0}{n_2 \pi}\right)^{1/2} \exp\left[-\left(\frac{r}{w}\right)^2\right]$$

Remarks

The agreement between the exact field and the field given by the Gaussian approximation is fairly good in the cutoff wavelength vicinity.

The power carried by the electromagnetic field is:

$$1/2 \int_s E \wedge H \, ds$$

This power does not depend on Z_0/n_2.

For the power P_c flowing through the core compared to the total power P_t, we obtain for LP_{01} according to [11]:

$$\frac{P_c}{P_t} = 1 - \left(\frac{u}{V}\right)^2 \left(1 - \left[\frac{K_0(\upsilon)}{K_1(\upsilon)}\right]^2\right)$$

and in Gaussian approximation:

$$\frac{P_c}{P_t} \approx 1 - \exp\left(-2\left(\frac{a}{w}\right)^2\right)$$

As expected, the approximation accuracy depends on $\lambda/\lambda c$. The accuracy is high near $\lambda/\lambda_c = 1$ but gives about 30% underestimation of P_c/P_t when $\lambda/\lambda_c = 2$. As the Gaussian approximation underestimates the evanescent field on the wavelength division multiplexers using dispersive elements such as gratings, the direct crosstalk given by calculations using the Gaussian approximation will generally be underestimated, as well as the width at the bases of spectral bandwidths (the crosstalk definition is given in Chapter 4 and the spectral widths of grating demultiplexers in Chapter 7). However, this difference remains very small for $\lambda/\lambda_c \sim 1$.

1.3 Useful approximations for single-mode fibres with graded index core

A graded index core fibre can be approximated to an equivalent step index core fibre. Then, the bandwidth and crosstalk analysis can use the Gaussian approximations of the equivalent step index fibre. Two single-mode graded index fibre types have been studied extensively.

Type 1

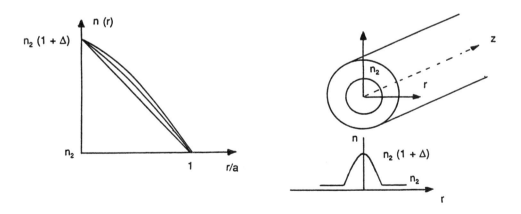

Figure 3
Structure and index profile of type-1 graded index fibre.

In the type-1 fibre, with index profile as shown in Figure 3, the core index $n(r)$ is given by the formula:

$$n^2(r) = n_2^2 \left(1 + \left(1 - \left(\frac{r}{a}\right)^g \right) 2\Delta \right) \quad \text{for} \quad 0 \leq \frac{r}{a} \leq 1$$

Type 2

In the type 2 fibre, with index profile as shown in Figure 4, the core index n(r) is given by the formula:

$$n^2(r) = n_2^2 \left[1 + \left(1 - \gamma \left(1 - \frac{r}{a} \right)^g \right) 2\Delta \right] \text{ for } 0 \leq \frac{r}{a} \leq 1$$

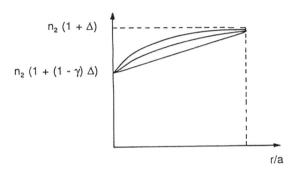

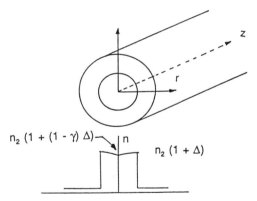

Figure 4
Structure and index profile of type-2 graded index fibre.

The propagation through graded index core fibre can be analyzed using elaborate theoretical calculations. For our purpose, it is generally sufficient to use the equivalent step index method applicable in a limited wavelength range:

$$0.8 \leq \lambda/\lambda_c \leq 1.5$$

An equivalent normalized frequency V_e, an equivalent index variation Δ_e and an equivalent core radius a_e, which give good approximations to the main propagation parameters of lower-order modes, are given in [11].

1.4 Remarks on dispersion properties

The propagation constant β is such that:

$$\beta^2 = k^2 n_1^2 - \frac{u^2}{a^2}$$

A normalized propagation constant is defined such that:

$$b(V) = \left(\frac{v}{V}\right)^2 = 1 - \left(\frac{u}{V}\right)^2$$

The index difference Δ between core and cladding is assumed to be small. ζ is the group delay per unit length, i.e. the reciprocal of group velocity V_g:

$$\zeta = \frac{d\beta}{d\omega} = \frac{1}{c}\frac{d\beta}{dk}$$

It was shown, [9], [10] and [11], that the group delay variation with wavelength λ, is:

$$\frac{d\zeta}{d\lambda} = -\frac{\lambda}{c}\frac{d^2 n_2}{d\lambda^2}\left(1 + \frac{d\,Vb}{dV}\Delta\right)$$
$$-\frac{n_2\Delta}{c\lambda}\left[\left(\frac{d\,(kn_2)}{dk}\frac{1}{n_2} - \frac{\lambda}{2\Delta}\frac{d\Delta}{d\lambda}\right)^2 V\frac{d^2 Vb}{dV^2}\right.$$
$$\left. -\frac{d\,(kn_2)}{dk}\frac{\lambda}{2n_2\Delta}\frac{d\Delta}{d\lambda}\left(\frac{3dVb}{dV} - b\right) + \lambda\frac{d\left(\frac{\lambda}{\Delta}\frac{d\Delta}{d\lambda}\right)}{d\lambda}\frac{b + \frac{dVb}{dV}}{2}\right]$$

The first term corresponds to the material dispersion in the wavelength range considered. The term proportional to the second derivative of V_b corresponds to the guide dispersion. The other part contains mixed terms. An approximated value for silica material dispersion is [12]:

$$M = 2.66\,10^{-2}\,\lambda - \frac{6.985\,10^{10}}{\lambda^3}$$

M in ps/(nm·km) and λ in nanometers. This expression does not vary very much with the fibre doping. $d\zeta/d\lambda$ decreases when Δ increases or when the cutoff wavelength decreases. The wavelength at which the chromatic dispersion is zero can be modified on silica fibres between 1.3 and 1.4 µm by doping variation.

More doping shifts the zero towards longer wavelengths. The limitation comes from the transmission losses related to higher dopings. Conventional unshifted single-mode fibres have zero-dispersion wavelengths near 1310 nm and typically a 16 to 18 ps/(nm·km) dispersion. The zero-dispersion wavelengths can be shifted near 1550 nm where the inherent loss of all silica-based single-mode fibres is also at a minimum [440].

Note
In pure silica, the material dispersion M goes down to zero at 1.27 µm. Germanium doping displaces this value towards longer wavelengths (λ = 1.35 µm) for a mole fraction of GeO_2 in silica of 13% [13]. For fluor doping, see [14].

1.5 Depressed Inner Cladding (DIC) fibres and other modified core fibres

The profile of such fibres is given in Figure 5.

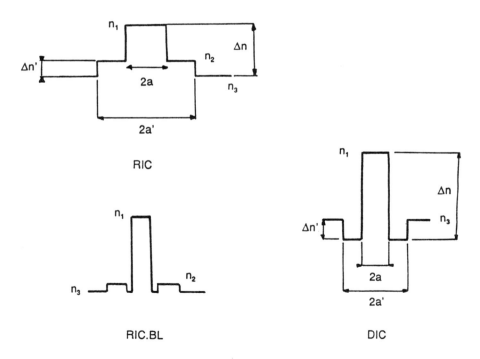

Figure 5
Different types of index profiles of fibres with modification of inner cladding index.

The modal dispersion depends on the curvature of the propagation constant as a function of the frequency $\omega = (2\pi/\lambda)$. As underlined in [15], the DIC structure allows a negative value of $V [d^2(Vb)/dV^2]$ to be obtained in single-mode propagation below $\lambda = 1.3$ µm that can compensate the silica negative dispersion in this wavelength range, or a positive value of $V [d^2(Vb)/dV^2]$ in single-mode propagation around $\lambda = 1.55$ µm that can compensate the silica positive dispersion about this wavelength, the latter case being more interesting because the losses at 1.55 µm are at a minimum. This fibre type is very important for wavelength division multiplexing. Practically, DIC fibres with a narrow inner cladding (typically $a'/a = 2$ and $0.5 \leq \Delta n'/\Delta n \leq 1$) could have dispersion lower than 1 ps/(nm·km) in the 1.34 - 1.58 µm spectral range, together with a low attenuation (0.26 dB/km at 1.52 µm), [15] and [16].

As in the case of graded index fibres, an equivalent step index fibre method can be used to calculate approximately the main propagation characteristics. In the DIC fibres, if $a'/a > (-\Delta n'/\Delta n)^{-1/2}$, the LP_{01} mode has a cutoff frequency different from 0. The approximation for this mode will be valid only far enough from the cutoff wavelength. Generally, the method can be applied with $-1 < \Delta n'/\Delta n \leq 0.2$.

Such an analysis, as well as theoretical and practical considerations on modified core fibres, can be obtained from [9] to [21].

1.6 Soliton propagation

The group velocity V_g depends on the optical frequency in an optical fibre. With classical fibres, the dispersion $dV_g/d\lambda$ is negative in the wavelength range corresponding to lower losses (~ 0.16 dB/km, for example, near 1.55 µm with high-quality fibre). The fibre group velocity responsible for pulse broadening can be compensated for particular pulses that have a critical power and shape and are called 'solitons'. Such pulses are produced by a self-phase-shift induced by refraction index variations upon the field intensity, I (known as Kerr effect) with silica glass $n \simeq n_0 + 3.2 \times 10^{-16}$ I, with I in watt/cm^2 and in which n_0 is the index at an arbitrary low intensity I. The response time of the phenomenon comes from electronic structure distortion in less than 1 fs (10^{-15}s = 1 fs) and from nuclei motion in about 100 fs in silica.

In soliton propagation, the velocity of the trailing half of the pulse tends to be increased and the velocity of the leading half of the pulse tends to be decreased by the non-linear effect; this compensates the reverse effect due to fibre dispersion. Soliton propagation implies a particular power and pulse shape. With pulses of a few picoseconds' duration, the index variation is relatively fast compared to the pulse duration. A given power level is required to reach the soliton propagation condition. The fundamental soliton of peak power P propagates along a lossless fibre without shape change for an arbitrary long distance. At $N^2 \cdot P$ peak power, where N is an integer, we obtain higher-order (N) soliton propagation. The higher-order solitons have a complex behaviour: they undergo periodic shape variations along the fibre length but return to their original shape after a length Z_{s0} called the soliton period. With pulse width T_0:

$$Z_{s0} = \frac{\pi}{2} \frac{T_0^2}{|\beta_2|}$$

where $\beta_2 = \frac{\lambda^3}{2\pi c^2} \frac{d^2 n}{d\lambda^2}$ is the group velocity dispersion [22].

The first theoretical study of soliton propagation in fibre was carried out by A. Hasegawa and F. Tappert in 1973 [23]. However, an amazing phenomenon involving a wave which propagated without shape modification was reported by Russel, a Scottish naval architect in 1834! He had observed a slow-moving wave which propagated about two miles down a canal without losing its shape. He called it 'solitary wave'. Since then, many researchers have contributed to the study of such propagation. The contribution of Bell Labs to this research has been important, see, for instance, [24] [25]. Coupled with wavelength division multiplexing, this propagation mode is very promising for ultra long-distance transmission; 1 million kilometres was demonstrated in 1992! [26]. The same year, NTT demonstrated 20 Gbit/s soliton transmission over 1020 km, using an erbium doped amplifier and forecast Tbit/s transmission with wavelength division multiplexing, and a 70 Gbit/s fibre based source of fundamental solitons at 1550 nm was designed by scientists from Russia and England [443].

As stated by L. F. Mollenauer and K. Smith at ECOC 1989: 'Clearly, the soliton is the natural and the only truly suitable pulse for a long-distance, all-optical communication system' [27].

> *The impatient reader, perhaps, is by this time accusing me of keeping the sun from him with a candle.*
>
> Sir Walter Scott

CHAPTER 2

Multimode fibres: geometrical optics analysis

We limit ourselves to considerations that will be useful later in this book for multimode fibre bandwidth calculations. Our main interest is the light propagation in multimode graded index fibres in which the index continuously varies as a function of the lateral distance from the axis. Let us recall that when the index variations are slow enough and when the fibre diameter $\varnothing$ is sufficiently large as compared to wavelength (for example $\varnothing = 50$ µm for $\lambda = 1$ µm), the laws of geometrical optics can be applied to the calculation of light ray trajectories, the wavefront being approximated locally by its tangent plane.

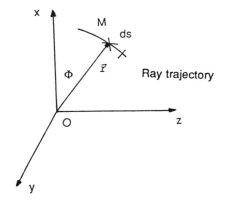

Figure 6a
Geometrical optics analysis - Notations.

Referring to Figure 6a, $M(\vec{r})$ is a point on the trajectory S, O is an arbitrarily defined origin. $\dfrac{d}{ds}\left(n\dfrac{\delta \vec{r}}{\delta s}\right) = \vec{\nabla}(n(\vec{r}))$ is the ray equation deduced from the Eikonal equation (see for example [28]). The fibre structure being defined (index at any point, geometry), then the propagation conditions are given by this equation to which it is necessary to add the incidence conditions on the entrance face. For graded index fibre it may be useful to define two new $\overline{E}$ and $\overline{I}$ variables:

$$\overline{E} = n\left(\dfrac{\delta z}{\delta s}\right) \qquad \overline{I} = nr^2\left(\dfrac{\delta \phi}{\delta s}\right)$$

($\overline{E}$ in analogy with energy and $\overline{I}$ in analogy with the angular moment of particle movement in a radial force field). $\overline{E}$ and $\overline{I}$ are given by the injection conditions. We can rewrite the ray equation using $\overline{E}$ and $\overline{I}$ [29]:

$$\dfrac{\delta^2 r}{\delta z^2} - \dfrac{\overline{I}^2}{r^3} = \dfrac{1}{2\overline{E}^2}\dfrac{\delta(n^2)}{\delta r}$$

Developing this analysis, it can be shown that the rays, injected at M_0 in the plane containing the z axis and M_0, will remain in this plane and will have an oscillatory movement.

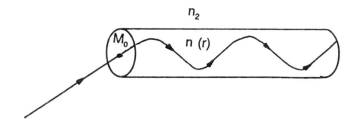

Figure 6b
Trajectory of rays injected in the (M_0,z) plane

The rays injected at M_0 at an angle with the plane containing z and M_0 will have a corkscrew trajectory in the z direction.

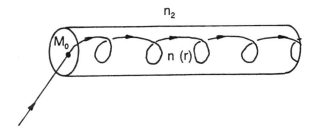

Figure 6c
Trajectory of rays injected at an angle with the (M_0, z) plane.

It can easily be shown that at each M_0 point of the injection section, there exists a local ray acceptance cone with a half-angle θ such that:

$$\sin\theta = \sqrt{n^2(r) - n_2^2}$$

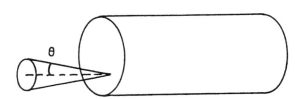

Figure 6d
Local numerical aperture - Notation.

So the numerical aperture decreases from the centre of the entrance section where it is maximum to the core edge where it goes to zero. An angle domain in which the rays are guided with leaks exists (in black on Figure 6e). We will not consider this leaky guiding angular domain in our bandwidth calculations, as the corresponding trapped energy is attenuated progressively within relatively small distances.

In the step index fibre case, the propagation angle is preserved at absolute value. The trajectory is a sequence of straight lines. The local numerical aperture is:

$$\sin\theta = \sqrt{n_1^2 - n_2^2}$$

which is the same over the whole core region.

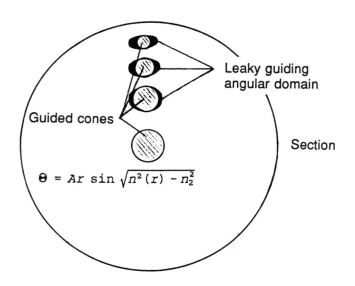

Figure 6e
Injection cones in a graded index fibre core depending on the entrance location.

Einen Regenbogen, der eine Viertelstunde steht, sieht man nicht mehr an.[1]

Goethe

CHAPTER 3

Wavelength division multiplexing: basic principles

It is possible to couple sources emitting at different wavelengths, $\lambda_1, \lambda_2, \lambda_j ... \lambda_n$, into the same optical fibre. After transmission on the fibre, the $\lambda_1, \lambda_2, ... \lambda_n$ signals can be separated towards different detectors at the fibre extremity (Figure 7a). The component at the entrance must inject the signals coming from the different sources into the fibre with minimum losses: this is the multiplexer. The component separating the wavelengths is the demultiplexer. The multiplexer may be replaced by a simple optical coupler, but division losses would then be increased. Obviously, when the light propagation is reversed, the multiplexer becomes the demultiplexer, and conversely. However, the coupling efficiency is not necessarily preserved in this operation. For example, if the multiplexer uses single-mode entrance fibres and a multimode output fibre, the coupling losses would be excessive in the reversed use. Multiplexers designed with identical input and output fibres are generally reversible. Simultaneous multiplexing of input channels and demultiplexing of output channels can be performed by the same component: the multi/demultiplexer. When a multiplexer has only two channels, it is naturally called a duplexer.

It should be noted that sources emitting simultaneously different wavelengths with close-set emitting areas are available; in this case, the multiplexer is unnecessary. The use of a receiver with several superimposed independent elements that are sensitive to different wavelengths is likewise proposed. Use of these solutions is often referred to as 'active multiplexing', meaning that they only use active microelectronic devices as multi/demultiplexers.

---

[1] If a rainbow lasts more than a quarter of an hour, one ceases looking at it

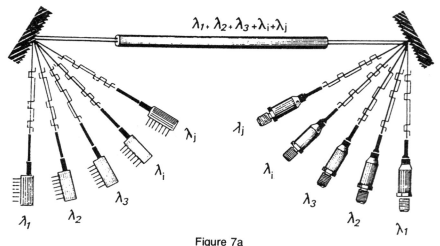

Figure 7a
Wavelength division multiplexing.

With modern commercially available telecommunication fibres, it is possible to transmit information over a large spectral range (Figure 7b). On multimode silica VAD[2] fibre without OH impurity, losses lower than 4 dB on 2.4 km were obtained from 0.65 to 1.9 µm as long ago as 1982 [30]. Now, single-mode silica fibres with minimum loss of about 0.16 dB/km at 1.55 µm are available. On multimode fibres, a graded index design allows a temporal dispersion minimization, but the optimum profile very much depends on wavelength and material. Thus, the bandwidth varies with wavelength. This problem is solved in part on 'double window' fibres. For example, 400-800 MHz·km bandpass at about 0.85 and 1.3 µm, with 2.4 dB at 0.85 µm and 1 dB at 1.3 µm losses, can be obtained [31].

With single-mode fibres for long-distance transmission and high bit rates (100 to 200 km with 10 GHz·km), dispersion lower than 1 ps/(nm.km) can be obtained between 1.34 and 1.58 µm by the MCVD[3] process.

With multiple cladding structures, dispersion smaller than a few ps/(nm.km) between 1.3 and 1.7 µm can be obtained, the losses being as small as 0.45 dB/km at 1.6 µm and 0.55 dB/km at 1.3 µm [28]. (For modified core fibre, see Section 1.5, Chapter 1). For dispersion properties of depressed inner cladding fibre, see [9].

Depressed cladding fibres with comparable attenuations and 14 ps/(nm.km) dispersion at 1.5 µm have been used in multiplexed systems at two wavelengths: 1.3 and 1.5 µm [32].

---

[2] VAD: Vapour phase axial deposition.

[3] MCVD: Modified chemical vapour deposition.

Materials other than silica are already being used in applications such as telespectrometry, laser beam energy conveyance and biomedicine. In the future, their use may increase due to their low loss and spectral range, especially towards the longer wavelengths [33]. Materials such as fluoride glasses are among the best candidates. In 1992, very high purity materials were obtained in a special crucible with a thin cushion of gas preventing the material from touching the inner wall of the crucible by CEREM S.A. (France). This was claimed to give fluoride glass much better transparency than that of silica. In such fibres, Rayleigh scattering which corresponds to a fundamental limit of the minimum loss would be much reduced by operating at longer wavelengths. In some short-distance applications, the use of plastic fibres becomes more and more interesting. In such cases, the wavelengths preferred are in the visible range. In 1992, graded-index and single-mode polymer optical fibres with a bandwidth of 2 GHz·km were obtained by the interfacial-gel polymerization technique. The transmission minimum attenuation was 56 dB/km at the 688 nm wavelength and 94 dB/km at the 780 nm wavelength [441].

The main sources are light emitting diodes (LEDs) or laser diodes (LDs). LEDs with maximum at 660 nm (GaAsP), between 740 and 950 nm (GaAlAs), between 1180 nm and 1340 nm (InGaAsP) and between 1500 and 1570 nm (InGaAsP) are available commercially. By 1992, LEDs with a typical coupling to a 62.5 µm fibre core, of 35 µW and 600 MHz bandwidth at 1320 nm were already available commercially. LDs that can be used at ambient temperatures in continuous waves (cws) are also available: between 630 and 680 nm (InGaAsP), between 740 and 905 nm (AlGaAs), between 910 and 990 nm (InGaAs), between 1180 and 1330 nm and between 1300 and 1600 nm (InGaAsP).

In 1992, a power of 2 mW at 625 nm, 22 °C, cw, was obtained in InGaP/InGaAlP [34]. Some LDs are tuneable over a relatively large spectral range (for example 1280 to 1330 nm or 1500 to 1565 nm [35]). Miniature Nd Yag lasers (laser diode pumped) can produce hundreds of milliwatts. However, they need external modulation, but they have several lines for multiplexing in the 1300 and 1064 nm range. The bandwidth of LDs can be very high. Thus, InGaAsP distributed feedback laser diodes (DFB lasers such as $\lambda/4$ 'shifted' 1.5 µm DFB-DC-PBH with a narrow MESA structure) at 8 GHz were reported in 1988 [36]. And in 1991, ATT-Bell Laboratories' scientists designed a monolithic InP-based laser in which pulses were generated at a 350 GHz repetition rate.

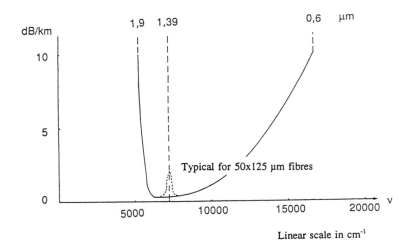

Figure 7b
Typical loss of a low OH content fibre.

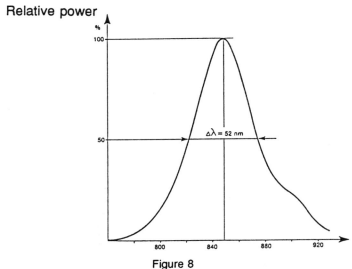

Figure 8
Typical spectrum of an LED.
(Thomson CSF.)

When the spectrum scale is given in wavenumber σ (cm^{-1}), which is, by definition, the wavelength inverse $\sigma = 1/\lambda$, the spectral width of LEDs does not vary very much over the usual telecommunication range (this is not true when the spectral width is given in wavelength units). The spectral width of a LED is typically: $\Delta\sigma = 600$ cm^{-1} $\pm$ 120 (Figure 8) [37]. The channel spacing is often chosen to be about 2 $\Delta\sigma$, but it can be slightly smaller. For example, with a typical LED, a spectral distance of 1100 cm^{-1} between the channels gives small emission spectral overlaps: $I_i(\lambda)$ and $I_j(\lambda)$ being the emission spectra of two adjacent channels, we obtain in the overlapping domain: $I_i(\lambda) < 0.1\ I_i$ maximum and $I_j(\lambda) < 0.1\ I_j$ maximum. In this way, we can theoretically have nine channels in the total silica fibre wavelength range, but in practice, this number is limited to five or six with the LEDs commercially available in 1992. However, this number can be greatly increased for short-distance, low-bit-rate links in which it is possible to tolerate some spectral overlap, or even to multiplex signals coming from identical LEDs (see LED slicing in Chapter 11, also [38] and [39]).

The spectral width of multimode lasers is generally smaller than 30 cm^{-1} corresponding to $\Delta\lambda = 5$ nm at $\lambda = 1300$ nm; for example, see Figure 9).

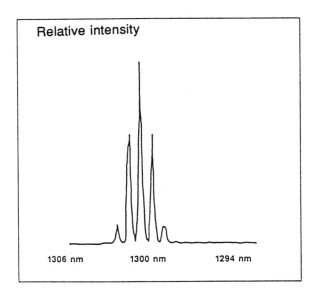

Figure 9
Typical emission spectrum of a multimode laser.
(Tektronics.)

Spectral widths lower than $1.7\ 10^{-6}$ cm^{-1} can be obtained with single-mode lasers. We conclude that, theoretically, a huge number of channels can be implemented. However, we are far from that possibility when practical problems with commercially available lasers are considered. First, a large manufacturing wavelength spread must be taken into account: at 1.3 µm, for instance, a 2 nm (12 cm^{-1}) spread on the same wafer and 10 nm (59 cm^{-1}) between wafers for buried heterostructure lasers [40]. The manufacturers give even larger tolerances. On the other hand, thermal drift is also a limiting factor (for instance, 0.1-0.35 nm/°C between 820 and 850 nm, corresponding to 1.43–5 cm^{-1}/°C). The problem at 1.3 or 1.5 µm is that it is necessary to control the temperature as the throughput depends too much on it. The wavelength drift with the laser electric current intensity must also be considered (see Section 15.7). Moreover, variations of λ may also be caused by feedback effects coming from selective chromatic reflections from different elements along the transmission line (connectors, couplers, commutators, etc.). But this last effect can be used for wavelength stabilization or monitoring in special cases. A typical emission spectrum of a single-mode laser is shown in Figure 10.

In 1992, at CLEO Conference, Glance *et al* described a broadband optical wavelength shifter using a semiconductor optical amplifier based device on which the output could be tuned over tens of nanometers, thereby increasing the number of channels allowed in WDM networks.

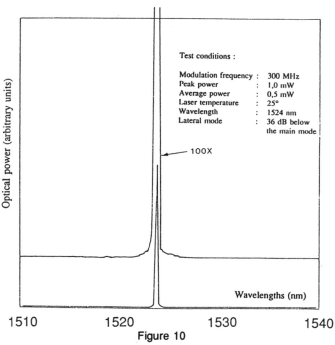

Figure 10
Typical emission spectrum of a single-mode laser.
(Lasertron.)

*Will Licht einem Körper sich vermählen,
Es wird den ganz durchsichtgaten wählen.*[1]

Goethe

CHAPTER 4

Multiplexers/demultiplexers: main characteristics

The multiplexer must combine the signals with minimal losses. Those losses P_j are expressed in decibels (dB) at each wavelength λ_j by:

$$P_j = 10 \log \left(\frac{\Phi_j}{\Phi_0} \right)$$

where Φ_j is the optical power injected into the transmission line and Φ_0 is the incident power at λ_j [41].

At the other end of the fibre, the signals at the different wavelengths are separated by a demultiplexer which, like the multiplexer, must have minimal losses. The optical crosstalk D_{ij} of a channel i on a channel j is:

$$D_{ij} = 10 \log \left(\frac{\Phi_{ij}}{\Phi_{jj}} \right)$$

where Φ_{ij} is the residual optical power of channel i at wavelength λ_i in channel j and Φ_{jj} the exit optical power in channel j at wavelength λ_j.

The total optical crosstalk in channel j is:

$$D_j = 10 \log \left(\frac{\sum_{i \neq j} \Phi_{ij}}{\Phi_{jj}} \right)$$

---

[1] When light seeks union with a body, it will choose one which is completely transparent.

This defect is merely due to the demultiplexer when sources with spectral widths much smaller than the multiplexer spectral passbands are used. But it also becomes necessary to take into account the multiplexer crosstalk in other cases. In such cases (such as LED slicing, for example – see Chapter 11) the optical crosstalk is a complex function of sources, multiplexer and demultiplexer.

The electrical crosstalk also depends on the receivers. At equivalent receiver sensitivity, in general, electrical crosstalk is twice as small in dB as is the optical crosstalk. The electrical crosstalk of a system also depends on the relative power and spectral width of the emitters, on the fibre spectral transmission and on the receiver's sensitivity variation with wavelength. For example, the difference is about 50 dB in a silicon receiver between 0.8 μm (high sensitivity) and 1.3 μm (almost no sensitivity).

When the same optical component performs multiplexing of some wavelengths and demultiplexing of the other wavelengths, we use a similar crosstalk definition, but the terminology 'near end crosstalk' is used for the parasitic effect of sources on receivers in the vicinity of these sources. The crosstalk coming from sources located at the other end of the line is called 'far end crosstalk'. It is obvious that the intrinsic near end crosstalk of a component must be several orders of magnitude lower than its far end crosstalk specification because the noise of sources not attenuated by the transmission line is superimposed through the component crosstalk to the signal of a source attenuated in the transmission line. Components with intrinsic near end optical crosstalk lower than – 120 dB have been manufactured (see Trimax, Figure 19, Chapter 6). The near end crosstalk also depends on the different reflections along the line, in particular on the number of connectors and on their location along the transmission line.

Of course, as with the products of all creative abilities, there is a real sense of beauty associated with a good design...

Angus Macleod

CHAPTER 5

Multidielectric filters used in multiplexing

Multidielectric filters produce a light beam angular separation in reflecting a given spectral range and transmitting the complementary part. These two spectral ranges can be very large, particularly with current edge filters: long wavelength pass filter LWPF and short wavelength pass filter SWPF. These filters consist of stacks of alternately high (H) and low (L) index layers on a substrate (S). Each layer has an optical thickness such that ne = $\lambda_0/4$ in order 0 filters and ne = $3\lambda_0/4$ in order 1 filters. Often, stack structures (H/2 L H/2)^K [37] are used and the main problem is to obtain sharp edge and high reflectivity power filters (R > 99%) in a given spectral range, and simultaneously obtain a good transmission in the complementary spectral range (T > 99%). If the (H/2 L H/2) elementary sequence is rigorously repetitive, we obtain reflection curves such as those shown in Figures 11 and 12, with oscillations on edges that can be corrected by an admittance adaptation on the first and last few layers; unfortunately this is detrimental to the relative sharpness of the transmission edge.

In Figure 13, LWPF theoretical and experimental transmission curves (31 layers), in the 'first window' for 0.82/0.85 µm separation, of a filter consisting of a stack of modified thickness layers are presented. In Figure 14, the same curves and the theoretical curve of an equivalent filter with non-modified layers are shown for 1.3/1.5 µm separation.

The manufacturing of SWPF is more delicate, as order 1 filters (ne = $3 \lambda_0/4$) with corrected entrance and exit layers are again required, and, consequently, the total coating is thicker. In any case, stringent control rules must be applied with longer deposition durations in the SWPF case. In practice, a spectral separation,

$\Delta\lambda = 0,02$ λ between channels can be obtained, but to take into account possible manufacturing defects, we would prefer to recommend a larger spectral separation, $\Delta\lambda > 0,05$ λ. It must also be pointed out that the thermal stability of such filters is 3×10^{-2} to 10^{-1} nm/°C.

A third filter type is often used in systems in which the multiplexing (or demultiplexing) is carried out through successive injections (or extractions) by transmission of a thin spectral band and reflection of the complementary spectral domain. It consists of multiple cavity filters used as bandpass filters (BPF). In practice, it is possible to manufacture bandpass widths $\Delta\lambda/\lambda = 0.045$ with a stack of 23 layers and three cavities with TiO_2 and SiO_2, respectively high and low index layers [37], [42], [43] and [44] (Figure 15).

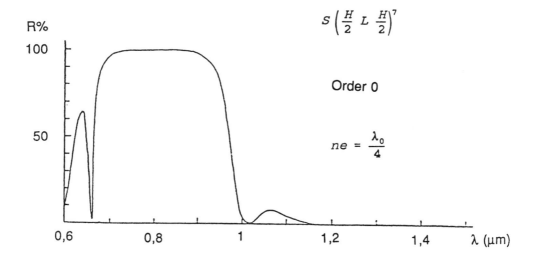

Figure 11
LWPF theoretical reflection curve without adaptation.

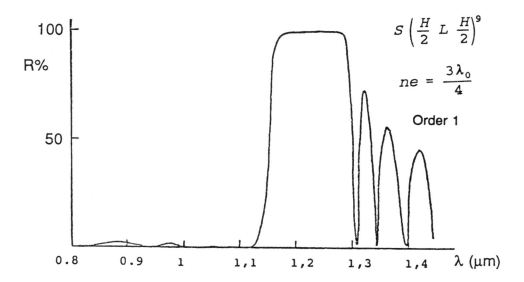

Figure 12
SWPF theoretical reflection curve without adaptation.

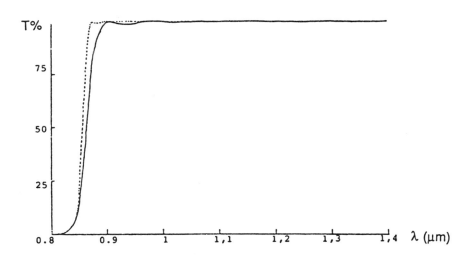

Figure 13
LWPF theoretical - - - and experimental ——— curves $(H/2\ L\ H/2)^{15}$ with adaptation.

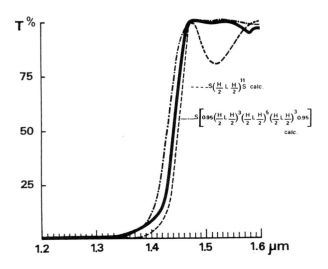

Figure 14
*1.3/1.5 μm filter transmission curves, theoretical without adaptation ---,
theoretical with adaptation -.-.-., experimental with adaptation ———.*

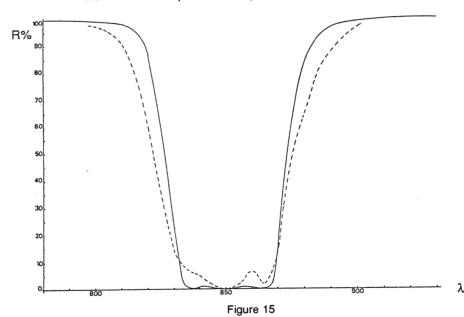

Figure 15
*Theoretical --- and experimental ——— curves
of a 23-layer multicavity BP filter
(high index 2.45, low index 1.47, substrate index 1.563).*

La conscience est originairement non pas un 'je pense que' mais un 'je peux'[1]

Merleau-Ponty

CHAPTER 6

Multidielectric filter devices

We will consider two types of multidielectric filter devices. In the first type, a direct coupling between fibres is used; in the second type the coupling between fibres uses focusing optics. We will give some examples.

The set up of Figure 16, from Trimmel *et al* [45], couples the fibres directly through dichroic filters integrated to the interfaces between the horizontal fibres (see the magnified drawing, inside circle, on Figure 16). This is a four-channel demultiplexer, it uses three filters of order 1 with modified layers:

$$S\ 1.015 \left(\frac{H}{2} L \frac{H}{2}\right)^2 \left(\frac{H}{2} L \frac{H}{2}\right)^9 1.015 \left(\frac{H}{2} L \frac{H}{2}\right)^2 S$$

The three filters are cascaded, they transmit λ_1 and successively reflect λ_2, λ_3 and λ_4. With single-mode fibres, the insertion losses of the component vary from 0.5 dB to 2 dB. On the same principle, components at 1.31 and 1.53 µm from Okuno and Kobayashi seem to give excellent results [46]. Among the devices using coupling optics, those with graded index lenses were introduced in 1977, to our knowledge [47]. An example is given in Figure 17 [48].

---

[1] Consciousness is not initially 'I think that' but 'I can'.

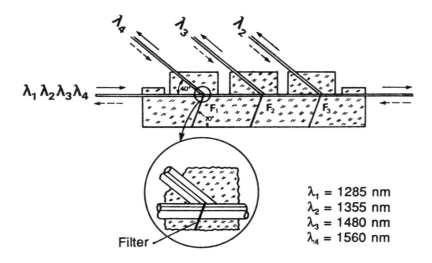

$\lambda_1 = 1285$ nm
$\lambda_2 = 1355$ nm
$\lambda_3 = 1480$ nm
$\lambda_4 = 1560$ nm

Figure 16
Single-mode fibres multidielectric filter multi/demultiplexer (from Trimmel and Malhein) [45].

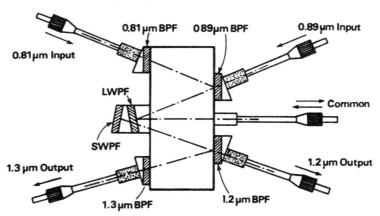

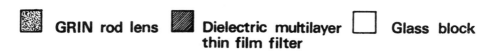

Figure 17
Four-channel multidielectric filter multi/demultiplexer using graded index lens coupling [48].

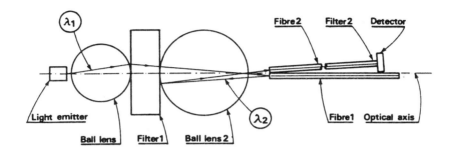

Figure 18
Ball lens duplexer (from Hillerich et al) [49].

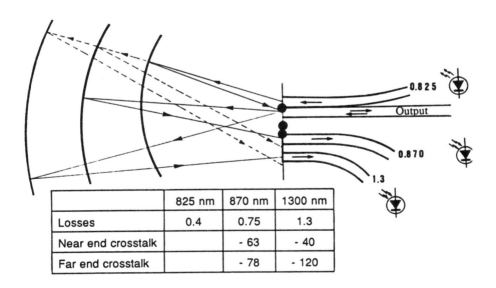

Figure 19
Trimax multi/demultiplexer [50].

In the component shown in Figure 17, the light beam is reflected between two parallel planes on which are located multidielectric filters. The light coupled to the line is separated into two spectral ranges: by a LWPF filter which reflects at 0.81 and 0.89 µm and by a SWPF filter which reflects at 1.2 and 1.3 µm. BPF filters separate the chosen channels: 0.81 µm is separated from 0.89 and 1.2 µm from 1.3 µm. The channel widths are 25 and 32 nm in the first window and 47 and 50 nm in the second window. The total losses are 1.4, 2.6 and 2.2 dB on the three longer wavelengths for the multiplexer and demultiplexer coupled with laser diodes, and 5.2 dB at 0.81 µm, this last channel using a LED. At the lower wavelength, the crosstalk is − 18 dB with the LED but only − 39 dB if a laser is used.

In the full duplex case, one wavelength must be injected and one wavelength must be coupled out. Hillerich *et al* set up [49] (Figure 18) is very simple. It uses ball lenses and two filters allowing 0.8/1.3 µm demultiplexing with a near end crosstalk of − 20 dB (for the definition of near end crosstalk, see Chapter 4).

We give another example of a bidirectional coupler for three wavelengths at 0.825, 0.870 and 1.3 µm: the Trimax [50] (Figure 19). In this component, the lower wavelength is reflected by the first mirror to the output. The two other mirrors separate 0.87 from 1.3 µm, reflecting them to fibres with detectors. The losses and the crosstalks of such a component are given in the table included in Figure 19.

Figures 20 − 22 show other multidielectric multiplexers or demultiplexers. The first uses graded index lenses and the reflections of light on two parallel planes [51]. The second also uses graded index lenses, but each filter wavelength can be adjusted by the rotation of dihedral mirrors at 90° without misalignment. That makes for a rather complex device but allows the use of lower tolerances on filter bandpasses [37].

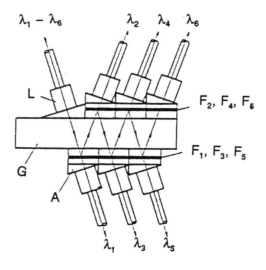

Figure 20
NTT 1978.

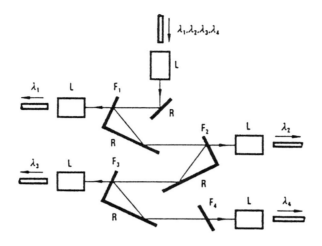

Figure 21
NEC 1980 [51].

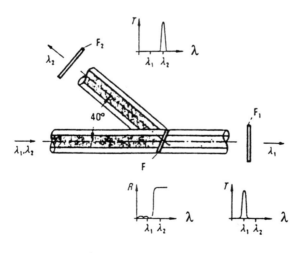

Figure 22
Siemens [53].

The set up of Figure 22 [53] is obtained with fibres joined as in Figure 16. An edge filter F and two BPF, F_1 and F_2 are used. Such devices can be manufactured as proposed by the authors [54], as shown in Figure 23.

The multiplexer of Figure 24 uses ball lenses and reflections between parallel planes [55]. The ball lenses focus the parallel beams coming from and to the plate on single-mode fibres (core diameter: 9 µm, cladding diameter: 125 µm).

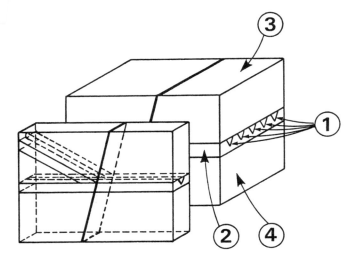

Figure 23
Collective manufacturing of multiplexers (described in Figure 22)
① V grooves - ② Resins - ③ and ④ Glass.

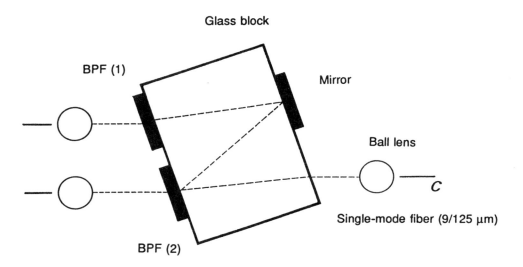

Figure 24
Device using ball lenses and reflections between parallel planes.

Si nous pouvions nous dépouiller de tout orgueil [...], nous ne dirions peut-être pas homo sapiens mais homo faber[1]

Henri Bergson

CHAPTER 7

Gratings for multiplexing

7.1 Introduction

Wavelength division multiplexers using filters cannot be used when the number of channels is too high or when the wavelengths are too close. The main advantage of the grating is the simultaneous diffraction of all wavelengths and so it is possible to construct simple devices with a very large number of channels.

A grating ([56] to [63]) is an optical surface which transmits or reflects light and on which a large number of grooves (several tens to several thousands per millimetre) are obtained by using a diamond tool or by holographic photoetching. The grating has the property of diffracting light in a direction related to its wavelength (Figure 25). Hence an incident beam with several wavelengths is angularly separated in different directions. Conversely, several wavelengths λ_1, λ_2, ... λ_n coming from different directions can be combined in the same direction. The diffraction angle depends on the groove spacing and on the incidence angle.

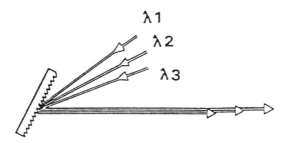

Figure 25
Principle of multiplexing by diffraction on an optical grating: wavelengths λ_1, λ_2, λ_3 coming from different directions are diffracted in the same direction into a single transmission fibre.

---

[1] If we could strip away our pride [...] we would perhaps not say homo sapiens but homo faber.

In Figure 26, let us consider a transparent and equidistant slit array and an incident plane wave at an angle with the perpendicular to the grating. Each slit diffracts light in transmission.

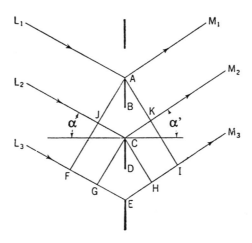

Figure 26
Calculation of the diffraction order angles.

Periodicity d

$$\Delta_0 = d\ (\sin \alpha + \sin \alpha') = k\lambda$$

Dispersion : $\dfrac{d\beta}{d\lambda} = \dfrac{k}{d \cos \alpha'}$

Resolution : $R_{max} = kN_0$

Orders position - Dispersion - Grating resolution.

In direction α', measured from the perpendicular to the grating, the waves coming from the different slits will be in phase if the path difference Δ_0 between the successive optical paths $(L_1\ M_1)\ (L_2\ M_2)$ is:

$$\Delta_0 = d\ (\sin \alpha + \sin \alpha') = k\lambda$$

where k is an integer, λ is the wavelength and d is the distance between two successive slits. $k = 0$ corresponds to direct transmission, $k = \pm 1$ corresponds to the first diffraction orders on each side of the direct transmission.

It is easily demonstrated that the angular dispersion, corresponding to the wavelength variation is:

$$\frac{d\alpha'}{d\lambda} = \frac{k}{d \cos \alpha'}$$

Consequently, the wavelengths can be separated angularly. The limit corresponds to the angular width of the diffraction of the whole surface of the grating projected in the α' direction. One can show that the maximum resolution that can be obtained is:

$$R_s \text{ max} = \frac{\lambda}{d\lambda} = kN_0$$

where N_0 is the total number of grooves. N_0 is usually very large, hence very small distances between channels can be obtained. In practice, spacings of 0.5 nm between channels have been obtained with grating multiplexers but the theoretical limit is far from being reached. In spectroscopy, resolution $\lambda/d\lambda = 0.5 \times 10^6$ is now standard with commonly used reflection gratings.

The reflection grating case is described in Figure 27. If the index outside the grating is n, the former law becomes:

$$nd (\sin \alpha + \sin \alpha') = k\lambda$$

in which d is the periodic distance between the grooves and α/α' the incident/ diffracted angles measured from the perpendicular N to the mean grating surface.

7.2 Efficiency versus wavelength

The groove shape allows the concentration of the diffracted energy in a given spectral range: the grating is 'blazed'.

7.2.1 Plane reflection gratings

N: perpendicular to the mean grating surface
M: perpendicular to the facet
α: incident angle (from N)
α': diffracted angle (from N)

i: incident angle (from M)
i': diffraction angle (from M)
d: groove spacing
γ: blaze angle

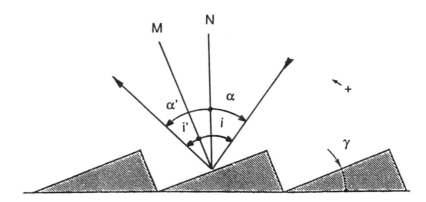

Figure 27
Diffraction plane grating.

A. Case in which the groove spacing is much larger than the wavelength and with small-angles, scalar approximation

We will use a small-angle approximation for α, α' and γ and assume that the number of grooves is large enough to have an angular width of diffraction by total grating surface much smaller than angular width of diffraction by facet.

We will obtain rough results but this analysis is very useful to start with.

Let us remark that the diffracted energy is maximum in the direction corresponding to a reflexion on each grating facet, i.e. when $i = -i'$.

From the relations $i = \alpha - \gamma$ and $i' = \alpha' - \gamma$, we obtain the blaze angle value

$$\gamma = \frac{\alpha + \alpha'}{2}.$$

This γ angle determines the shape of the diamond to be used for the ruling of the grating master.

The grating relation, $d(\sin \alpha + \sin \alpha') = k\lambda$ becomes:

$$2d \sin \gamma \cos \frac{(\alpha - \alpha')}{2} = k\lambda$$

For reflection gratings, the blaze angle is generally calculated in the Littrow conditions, in which $\alpha = \alpha'$, corresponding to an incident and exit beam in the same direction.

In the first order and Littrow condition:

$$\lambda_{1\,blazed} = 2d\sin\gamma$$

With $\alpha \neq \alpha'$:

$$\lambda_{1\,blazed} = 2d\sin\gamma\cos\frac{(\alpha - \alpha')}{2}$$

In the second order and Littrow condition:

$$\lambda_2 = d\sin\gamma$$

In the ξ order and Littrow condition:

$$\lambda_\xi = \frac{2d}{\xi}\sin\gamma$$

Each facet of the plane grating gives a diffraction phenomenon which is characterized by the distributed amplitude A:

$$A = A_{\lambda_\xi}\frac{\sin\frac{\pi d}{\lambda}(\sin i + \sin i')}{\frac{\pi d}{\lambda}(\sin i + \sin i')}$$

For $\lambda = \lambda_\xi$, the blaze wavelength, the intensity is maximum and equal to:

$$I_{\lambda_\xi} = \left(A_{\lambda_\xi}\right)^2$$

Let us calculate the ratio:

$$\frac{I_\lambda}{I_{\lambda_\xi}} = \Phi\left(\frac{\lambda_\xi}{\lambda}\right)$$

which represents the spectral distribution of the diffracted intensities. The intensity I_λ corresponding to the wavelength λ is:

$$I_\lambda = I_{\lambda_\xi}\left[\frac{\sin\left(k\pi - \frac{\pi\xi\lambda_\xi}{\lambda}\right)}{k\pi - \frac{\pi\xi\lambda_\xi}{\lambda}}\right]^2$$

in which for each wavelength $\alpha = \alpha'$, k is the diffracted order considered at λ and ξ is the diffracted order for which the grating is blazed at the wavelength λ_ξ.

Figure 28 shows the spectral distribution of the diffracted energy in the first order from a grating blazed in the first order ($\lambda_\xi = \lambda_1$).

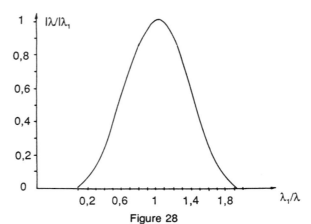

Figure 28
First-order Echelette grating efficiency (approximation of small blaze angle) (normalizated abscissa λ_1/λ).

B. General case

With gratings having a small groove spacing (d = a few λ or less), the results of section A are no longer valid. Anomalies appear in the efficiency curves and the spectral distribution of the diffracted energy depends on the polarization. The formula $\lambda_{1\ blazed} = 2\ d\ \sin \gamma$ is no longer valid. The maximum efficiency with unpolarized incident light is lower than λ_1 given by this formula. The corresponding curves were fully calculated from Maxwell's equations (M. Petit, Thesis, Faculté d'Orsay, France, 1966 and [56]). These curves give the efficiencies of perfectly conducting sinusoidal gratings (Figure 29a) or perfectly conducting triangular profile gratings (Figure 29b).

Efficiency versus wavelength

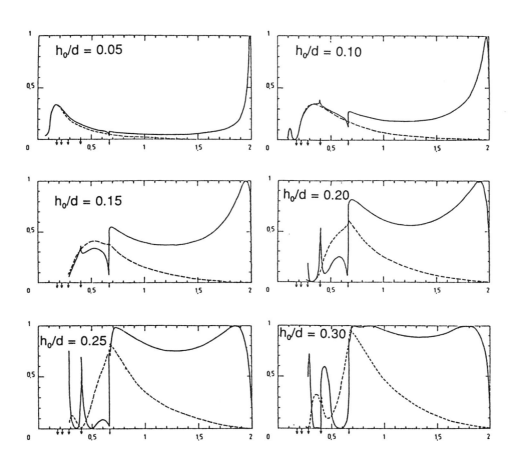

(—) TM polarization
(---) TE polarization

Figure 29a
Efficiencies as function of the ratio λ/d of sinusoidal profile reflexion gratings with different depths.
Electromagnetic theory [56] of perfect conductors.

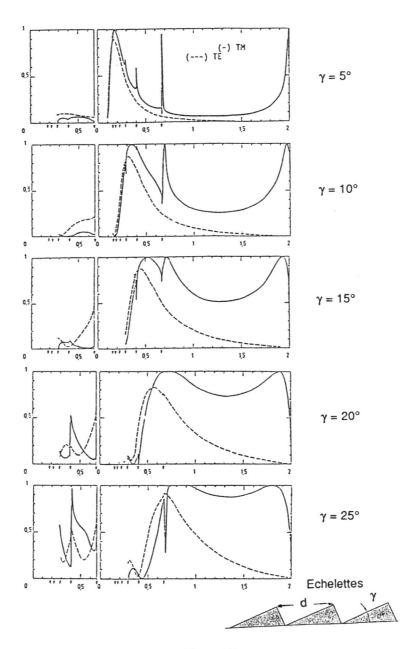

Figure 29b
Efficiencies versus wavelength/spacing ratio λ/d for different blaze angle triangular profiles γ between 5° and 25°.
Electromagnetic theory [56].

Efficiency versus wavelength

7.2.2 Transmission gratings

The grating grooves are transferred on to a resin coating on a glass blank with two faces polished to within a quarter of a fringe. The grooves can be considered as a set of small diffracting prisms.

Let us consider:
n resin index,
γ facet angle,
d grating period (see Figure 30a).

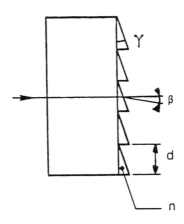

Figure 30a
Transmission grating.

A. Gratings with groove spacing much larger than one wavelength and small blaze angle (scalar approximation)

When the incident light is perpendicular to the blank (Figure 30a), the blaze wavelength λ_B is given by the formula:

$$\lambda_B = d\,(n - 1)\,\sin \gamma$$

B. General case

Here again, it is necessary to use the electromagnetic theory to obtain the correct efficiency value.

However, it can be demonstrated that one can obtain 100% efficiency by using the formula given above, with a metallic coating on the small facet of the grooves. This is obtained with an electromagnetic field perpendicular to the conductive facet [61].

7.2.3 Concave gratings

Concave gratings are generally used in reflection; therefore, their blaze angles are calculated like those of plane reflection gratings. In the scalar theory, the blaze angle of such gratings has to be changed continuously in order to keep constant the perpendicular to each facet, the bisector of ABC, the angle between incident and diffracted order 1, at all locations of B on the grating surface (Figure 30b). However, this is not usually necessary, and, in most cases, it is the angle between each facet and the plane tangent at the centre P which is kept constant. However, on ruled or holographic gratings, a profile variation with surface location can be obtained by variable incidence ion etching. On classically ruled gratings, one can rule three or four zones with a constant but optimized angle and with a corresponding loss of resolution, in cases where the efficiency variation from the centre to the edge would be too large. The resolution of such a grating ruled in three parts will be at least three times less than that of the same grating ruled as a single part, but, in most cases, it will have an efficiency approximately equal to the efficiency of the equivalent plane grating [62].

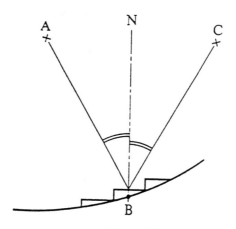

Figure 30b
Concave grating.

The possibility of controlling the focal properties of concave gratings by a proper distribution of the grooves has been known for a long time. This can be done on computerized ruling engines. However, it is more easily obtained using holographic techniques. During the last 15 years, holographic gratings have evolved dramatically. They are capable of stigmatic imaging without the need for auxiliary focusing optics. Ion etching provides a means of blazing and optimizing these gratings. Following production of the holographic master, which has pseudo-sinusoidal groove profiles, the grating is then used as a mask and subjected to an argon beam to remove surface atoms until the groove structure presented by the surface hologram is brought into the substrate itself. To 'shape' the grooves, the

Efficiency versus wavelength 47

angle of incidence of the ions to the substrate can be adjusted to produce triangular blazed grating profiles. High efficiencies have been achieved with concave aberration-corrected holographic gratings [468]. Typically, 60–80% efficiencies and subnanometer wavelength resolution from 1500 to 1560 nm are obtainable on all single-mode, concave holographic grating, tuneable demultiplexers.

7.2.4 Practical efficiency

Practically, with echelette gratings such as those of Figures 31a and 31b, 85% to 90% efficiency with unpolarized incident light is obtained at the blaze wavelength. Polarization by the grating remains small if the blaze angle is small enough (here < 5% on several tens nanometers). This efficiency will be increased by multidielectric coatings [63].

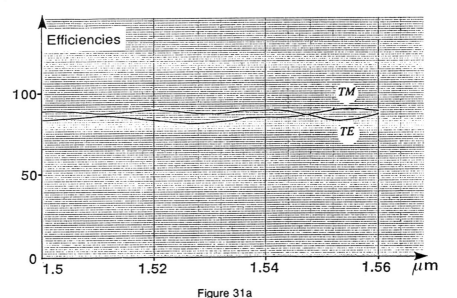

Figure 31a
TE and TM efficiencies measured on a small blaze angle grating (TE for transverse electric, TM for transverse magnetic).

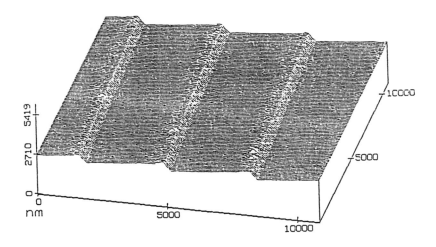

Figure 31b
Grating profile, tunnel microscope view.

7.3 Bandwidth of grating devices

7.3.1 Devices with input and output single-mode fibres

For the theoretical analysis, we use the general case of Figure 32 and calculate the transmission function from the entrance fibre F_1 towards the exit fibre F_2.

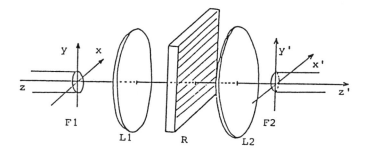

Figure 32
Theoretical device.

Bandwidth of grating devices

Let us consider a fibre F_1 with a polished end-face in the plane xy perpendicular to the direction zz' axis of the fibre. The focal plane of an optical system L_1 is on the xy plane. The end-face of F_1 is imaged in the focal plane of L_2, x'y', which contains the polished end-face of fibre F_2, after an angular dispersion of the dispersor R (generally a grating). xyz and x'y'z' are local right-angled trihedrals.

We will calculate the energy transmitted through F_2 as a function of the variation of the wavelength λ of the light delivered by L_2.

We assume hereafter that the magnification of the optical system is 1 and that the couple L_1 and L_2 is afocal. As a matter of fact, that condition is necessary to avoid a coupling loss (an afocal coupling with a unit magnification keeps an identity between entrance and exit angles. This is the case with the Stimax configuration (see Figure 45a), but it is not always verified with other configurations).

We show what happens in the x'y' plane in Figure 33.

The function A(x'y'), limited to S_1, corresponds to the incident amplitude in the x'y' plane. The function T(x'y') limited to S_2, corresponds to the amplitude transmission function of the fibre F_2. Therefore, the amplitude dA induced in F_2 for an elementary spectral width $d\lambda$ will be:

$$dA = A(x'y') \, T(x'y') \, d\lambda$$

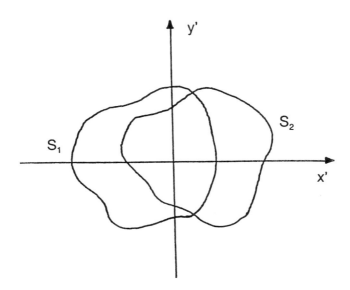

Figure 33
Transmission in the exit plane, general case.

When λ varies, the function A is translated in the plane $x'y'$.
If we assume that the dispersion is linear along x':

$$\lambda - \lambda_0 = \alpha (x' - x'_0)$$

This is only an approximation in which, for small angles, the sine of the diffraction angle is approximated to the angle.

The flux F will be the modulus square of the correlation of A with T:

$$F = \left| K \cdot \iint A(x' - x_0, y') \, T(x'y') \, dx' dy' \right|^2$$

Let us assume that, in the Gaussian approximation, with a mode radius ω'_0 corresponding to the half width of the amplitude distribution A at $1/e$:

$$A = \exp - \left[\frac{r'}{\omega'_0} \right]^2$$

with $r'^2 = x'^2 + y'^2$.

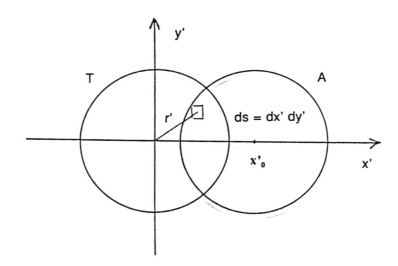

Figure 34a
Transmission in the exit plane, single-mode fibre case.

$$F = \left| K \int_{-\infty}^{+\infty} \int_{-\infty}^{+\infty} \exp\left[\frac{x'^2 + y'^2}{\omega_0'^2} \right] \exp\left[\frac{(x' - x'_0)^2 + y'^2}{\omega_0'^2} \right] dx' dy' \right|^2$$

From which:

$$F = \left| K \int_{-\infty}^{+\infty} \exp\left[\frac{2y'^2}{\omega_0'^2}\right] dy' \int_{-\infty}^{+\infty} \exp\left[\frac{x'^2}{\omega_0'^2}\right] \exp\left[\frac{(x'-x_0')^2}{\omega_0'^2}\right] dx' \right|^2$$

It can be seen that only the second integral depends on x_0'. Its direct calculation is relatively easy. But it is easier to consider this second integral as an auto-correlation function:

$$\left[\exp\left(\frac{x'^2}{\omega_0'^2}\right) \otimes \exp\left(\frac{x'^2}{\omega_0'^2}\right)\right]_{(x_0')} = [g(x') \otimes g(x')]_{(x_0')}$$

It is well known that $f(x)$ being a function and $\tilde{f}(u)$ being its Fourier transform, the Fourier transform of the function $g(x) = f(x/a)$ is $\tilde{g}(u) = |a| \tilde{f}(au)$ ('expansion' theorem), then:

$$g(x') = f(x'/\omega_0') = \exp-\left(\frac{x'}{\omega_0'}\right)^2$$

and

$$f(x') = \exp(-x'^2)$$

As the Fourier transform of $\exp(-\pi x'^2)$ is: $\tilde{f}(u) = \exp(-\pi u^2)$, the Fourier transform of $g(x')$ is:

$$\tilde{g}(u) = \sqrt{\pi} |\omega_0'| \exp(-\pi^2 \omega_0'^2 u^2)$$

In the case of symmetrical functions, the correlation and the convolution are identical. Moreover, it is known that the Fourier transform of the autoconvolution product of a function $g(x')$ is the modulus square of the Fourier transform of that function:

$$\mathrm{TF}\,[g(x') \otimes g(x')] = |\tilde{g}(u)|^2$$
$$|\tilde{g}(u)|^2 = \pi\omega_0'^2 \exp\left(-2\pi^2\omega_0'^2 u^2\right) = \pi\omega_0'^2 \exp\left(-\left(\pi\sqrt{2}\omega_0' u\right)^2\right)$$

Then, if we go back to the initial function through an inverse Fourier transform:

$$[g(x') \otimes g(x')] = \mathrm{TF}^{-1}\left[|\tilde{g}(u)|^2\right]$$

and if we again use the 'expansion' theorem on the inverse Fourier transform, we obtain:

$$[g(x') \otimes g(x')] = \sqrt{\pi}\,\frac{\omega_0'}{\sqrt{2}}\,\exp\left(-\frac{x'^2}{2\omega_0^2}\right)$$

or:

$$F = k\,\exp\left(-\left(\frac{x'}{\omega_0'}\right)^2\right)$$

Thus, the intensity transmission function $F(\lambda)$ is identical to the function representing the amplitude distribution $A(x')$ when the two functions are drawn on the same graph using abscissae $(\lambda - \lambda_0)/\alpha$ for $F(\lambda)$ and $x' - x'_0$ for $A(x')$.

The total width at half maximum of F is:

$$2\,\omega_0'\,\sqrt{\ln 2} = 1.6651\,\omega_0'$$

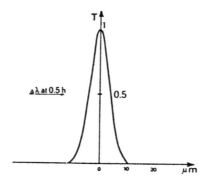

Figure 34b

Transmission versus distance between the exit fibre centre and the entrance fibre image centre.

7.3.2 Devices with a single-mode entrance fibre and an exit slit or a diode array with rectangular pixels

Let us consider Figure 35. We assume that the coupling device is similar to that of Figure 32, but an exit slit is placed at $x' = 0$ $y' = 0$, the height of the slit being parallel to the grating grooves and y'.

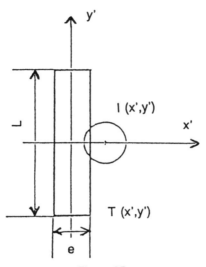

Figure 35
Exit slit and image of the entrance fibre in the exit plane.

We assume that the transmission $P(x',y')$ is uniform and equal to 1 inside the slit, which is considered to have a height much larger than the fibre core diameter, corresponding to an intensity distribution $I(x',y')$.

The elementary intensity dI transmitted through the slit for an elementary spectral width $d\lambda$ will be:

$$dI = I(x',y') \, T(x',y') \, d\lambda$$

As above, when λ varies, I is translated nearly linearly (same approximation as in Section 7.3.1). The transmitted flux is the modulus of the correlation of I with T:

$$F = K \cdot \iint I(x' - x'_0, y') \, T(x',y') \, dx' dy'$$
$$F = I \otimes T$$

The intensity distribution in the image (assumed perfect) of the single-mode fibre end-face is:

$$I = I_0 \exp - \frac{2 r^2}{\omega_0'^2}$$

The double integral can be decomposed into simple integrals on r. These integrals can be calculated using the Simpson method. We used only eleven elements, which correspond to a precision better than the measurement precision arising from experimental uncertainties. The maximum integration radius is 20 μm. We used slit widths of 5, 10, 12, 14 and 20 μm. Only for e = 5 μm is the integration method slightly different, and hence more precise. The results are given on the curves of Figures 37a for e = 5, 10, 12 and 20 μm, and 37b and 37c for e = 10, 12 and 14 μm compared to the results obtained with devices using input and output single-mode fibres.

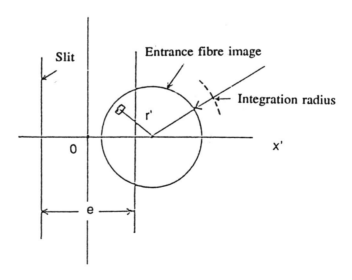

Figure 36
Notations in the exit plane.

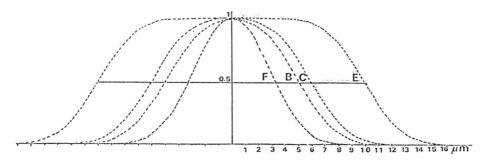

Figure 37a
Intensity transmission functions.
Single-mode fibre to exit slit.
Single-mode (SM) fibre, mode radius $\omega_0 = 4.573$ μm in slit F, slit width: e.

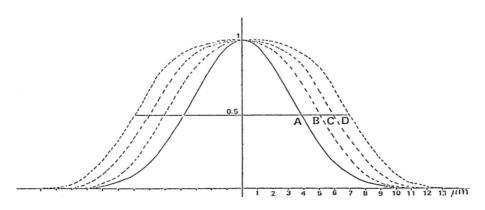

Figure 37b
Intensity transmission functions.
Single-mode (SM) fibre in, exit slit.
Single-mode (SM) fibre, mode radius $\omega_0 = 4.573$ μm
in an identical SM fibre and in slits with width e.

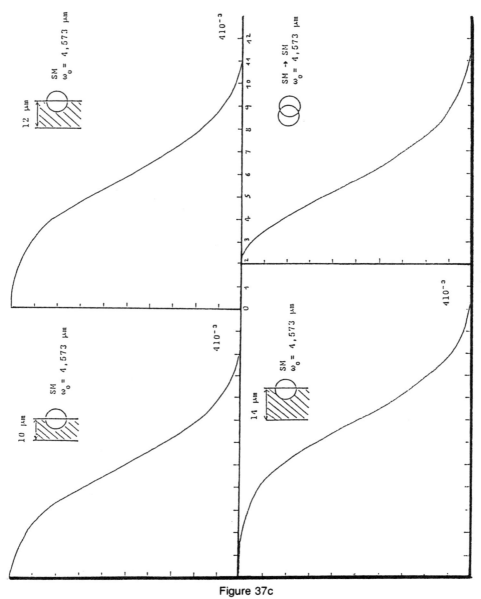

Figure 37c

Comparison of intensity transmission function of a single-mode fibre with a mode radius $\omega_0 = 4.573$ μm, through another single-mode fibre with a mode radius ω_0 (A), or through exit slits with different widths: 10 μm (B), 12 μm (C), 14 μm (D), 20 μm (E).

Intensity in the entrance fibre: $\quad I = I_0 \exp\left|\dfrac{-2r^2}{\omega_0^2}\right|$

7.3.3 Devices with a single-mode entrance fibre and multimode step index exit fibres

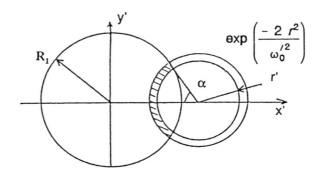

Figure 38
Multimode fibre core intercepting an SM fibre intensity distribution.

We consider that a multimode fibre with a numerical aperture that is always larger than the numerical aperture of the single-mode fibre collects all the available intensity in its core having a radius R_1. The integration is limited to a domain of radius R_2 much larger than ω_0' but with $R_2 < R_1$.

Let us consider an elementary surface $ds = 2\alpha\, r dr$. We must identify four cases:

1) $x_0' > R_1 + R_2$ $F = 0$

2) $R_1 < x_0' < R_1 + R_2$:

$$F = \int_{R_2}^{x_0'-R_1} 2r \exp\left(\frac{-2r^2}{\omega_0'^2}\right) \text{Arcos}\left(\frac{r^2 + x_0'^2 - R_1^2}{2 r x_0'}\right) dr$$

3) $R_1 - R_2 < x_0' \leq R_1$:

$$F = \int_{R_2}^{R_1-x_0'} 2r \exp\left(\frac{2r^2}{\omega_0'^2}\right) \text{Arcos}\left(\frac{r^2 + x_0'^2 - R_1^2}{2 r x_0'}\right) dr$$
$$+ \frac{\pi \omega_0'^2}{2}\left(1 - \exp\left(\frac{-2(r_1 - x_0')^2}{\omega_0'^2}\right)\right)$$

4) $x_0' < R_1 - R_2$:

$$F = \pi \frac{\omega_0'^2}{2} \left(1 - \exp\left(-\frac{2R_2^2}{\omega_0'^2}\right)\right)$$

We calculated these functions for the different cases (see Figure 39).

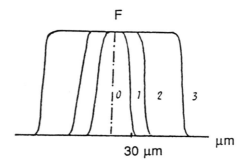

Figure 39
Entrance single-mode fibre - Exit step index fibre
Curves   1   2   3
R_1/ω_0' 9.09   18.18   36.36

7.3.4 Devices with a single-mode entrance fibre and multimode graded index exit fibres

A similar mathematical development leads to the following transmission curves:

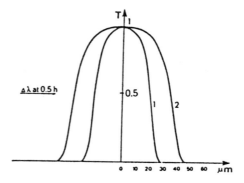

Figure 40
Single-mode entrance fibre and graded index exit fibre.

Curve	1	2
Entrance core diameter (in μm)	8	8
Exit core diameter (in μm)	50	85

7.3.5 Coupling in a device with small aberrations

In such a case, we mean that from a point located at the centre of the entrance fibre core, the device gives an image spot that is well focused to within a few micrometres.

Thus, the coupling losses are negligible in the single-mode to multimode case (usual demultiplexer: SM → 50 x 125 μm, for example) and the F function is not modified much.

This is not true when coupling between single-mode fibres (multiplexer) is considered.

In the entrance fibre image spot, the amplitude distribution A is modified by the aberrations. The new distribution A' will be the convolution product of A by the instrumental response P of the system to an elementary point:

$$A' = [A * P]_{(x'_0)}$$

We will obtain:

$$F = K \left| [A * P \otimes A]_{(x'_0)} \right|^2$$

The calculation is generally easier when performed in the exit pupil plane.

It can be demonstrated that the coupling losses remain smaller than 1 dB as long as the sum of the different aberrations measured in the exit pupil remains smaller than $\lambda/3$ [64].

> *One must be an inventor to read well*
> Ralph Waldo Emerson

CHAPTER 8

Grating microoptic devices

8.1 Grating multiplexers using all multimode fibres or grating demultiplexers with a single-mode entrance fibre and multimode exit fibres

Generally, grating multiplexers or demultiplexers consist of three main parts: entrance and exit elements (fibre array or transmission line fibre and emitters or receivers), focusing optics and dispersive grating. The grating is generally a plane grating [65] (Figure 41), [66] (Figure 42), [67] to [70] (Figure 43), [71] (Figure 44), [72] and [73] (Figure 45a).

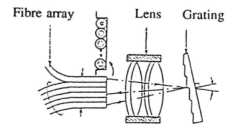

Figure 41
(Koh Tohii Aoyama et al.) [65].

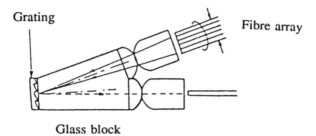

Figure 42
(Watanabe et al.) [66].

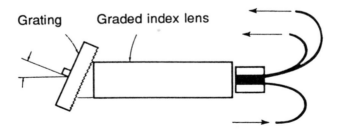

Figure 43
*Multiplexer with a plane grating and a graded index lens
(after Tomlinson) [67] and [70].*

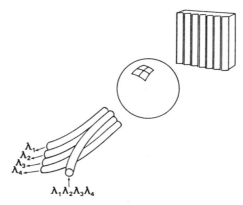

Figure 44
Grating multiplexer (after Finke et al.) [71].

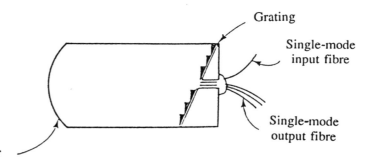

Figure 45a
*Multiplexer using single-mode or multimode fibres
(after J.-P. Laude) [72] (see also Figure 45b, p.64).*

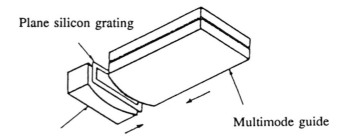

Figure 46
*Multiplexer using a flexible grating bent on to a multimode planar guide
(After Fujii et al.) [78].*

Grating

For instance, the Finke *et al* set up (Figure 44) is very simple in principle. The fibre array extremity is at the object focus of a ball lens, the grating being located at the image focus. The system is afocal, with magnification 1, so that all ray angles from and to the fibres are identical, as originally proposed for the configuration of Figure 45a. The authors claim 1.2–1.7 dB demultiplexer losses with four to six channels. Demultiplexers with up to ten channels were designed around such a configuration.

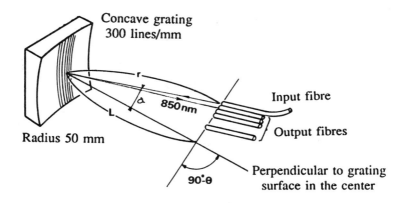

Figure 47
*Aberration-corrected grating multiplexer
(after Harada).*

As early as 1977, another configuration was proposed by Tomlinson (Figure 43) [67] and [70], and it is still often used. It consists of a graded index lens placed in front of a plane grating prism. In Europe, Mannschke obtained interesting results with a five-channel demultiplexer; losses were 0.9 to 2 dB [74] and [75]. In 1980, we published the configuration of Figure 45a: a fibre array is placed in front of a slit, photoetched on a plane reflection grating, perpendicular to the grooves. A concave mirror transforms the diverging beam coming from any fibre into a parallel beam; this beam coming to the grating is angularly dispersed back to the concave mirror and is imaged on the fibre array extremity in a position that depends on its wavelength. This configuration is aplanatic, afocal and has a magnification of 1. Thus, all angles from and to the fibres being identical, we obtain the best conditions for a high coupling efficiency. The achromatism is perfect and aberrations are almost nil when the mirror is parabolic. Indeed, a spherical mirror is often used because the aberrations remain very small. Depending on the available sources, components with up to six channels for LEDs and 22 channels for laser diodes are generally manufactured. However, for particular applications, multiplexed sensor components with up to 49 channels sliced from the same LED have been manufactured (Figure 45b). In each case, with monochromatic sources centred on the multiplexer transmission bands, the losses remain smaller than 2 dB and may reach 0.5 dB as in the 1.54/1.56 µm single-mode device of [76].

Figure 45b
49-channel grating demultiplexer for LED slicing (see Chapter 11)
(Laude et al, 1984).

The use of dedicated concave gratings [77], [78], [79] and [72] simplifies the device, as, for example, in the Harada *et al* configuration (Figure 47) in which 2.6 dB losses are reached on a four-channel multiplexer at 821.5, 841.3, 860.9 and 871 nm with 60/125 µm multimode fibres. However, it is impossible to retain aberration-free focusing over a large spectral range with such configurations. The devices from [80] and [81] are also interesting multiplexer examples from the early 1980s.

8.2 All single-mode fibre grating multi/demultiplexers

The multiplexer is always more difficult to manufacture than the demultiplexer because the core diameters are small and the highest optical quality is essential. This is particularly true when all single-mode fibre arrays are used. For example, if a λ/10 defect of spherical aberration, of coma and of astigmatism is allowed, the cumulative losses are 1 dB [64] and [82]. Moreover, in order to obtain a channel spectral width large enough, as compared with channel spacing, the geometrical distance between fibre cores must be small. The passbands are practically adjacent for a distance between fibres of 22 μm when 11 μm diameter core is used (Figure 48a). Fibres with a small cladding are necessary.

On this principle, Hegarty et al [83] manufactured a nine-channel single-mode multiplexer. The distance between fibres was 36 μm, the width of each channel was 0.2 nm at a wavelength about 1.5 μm, with 1.5 dB losses. The configuration of Figure 48b is often used for single-mode fibres. For cost saving, such an aberration correction is preferred here over the parabolic mirror solution (Figure 45a). With two materials having identical chromatic dispersion, the geometrical aberration can be corrected without introducing chromatism.

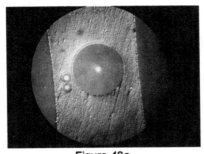

Figure 48a
Single-mode fibres at the focal plane: one 11/125 μm and two 11/20 μm, fibres spaced at 22 μm (Jobin Yvon documentation).

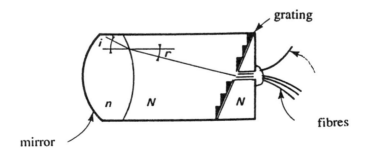

Figure 48b
*Single-mode fibre device
(aberrations 0.05 to 0.5 μm with ON = 0.2).
J.-P. Laude, 1984.*

The characteristics of some components using these techniques (Figure 45a or Figure 48b) are given in the following table, [84] and [85]:

Ref.	Δx (μm)	Focal length (mm)	dλ_FWHM (nm)	Δλ (nm)	Number of channels	λ (μm)
1	130	62.01	1.1	20	8	1.3 and 1.46 to 1.58
2	106	119.53	0.325	2	18	1.527 to 1.561
3	85	119.53	0.3	1.6	20	1.515 to 1.546
4	76	31.74	3.5	35	4	1.240 to 1.345
5	65	31.74	2.75	30	4	1.24 to 1.33
6	60.3	43.03	2.25	20	4	1.27 to 1.33
7	47.5	24.99	1.55	9	10	1.260 to 1.341
8	47.5	22.49	1.9	10	8	1.27 to 1.33
9	45	40.10	3.2	20	8	1.3 and 1.46 to 1.58
10	45	53.54	0.85	4	16	1.27 to 1.33
11	22	43.40	13	30	2	1.52/1.55
12	22	53.54	15	40	2	1.28/1.32
13			0.27	0.63	4	1.527 68 to 1.529 58

Δx is the distance between fibres in the focal plane and $\Delta \lambda$ is the distance between channels in nm. The losses vary from 1.5 to 4 dB. When the components are used as demultiplexers, the optical crosstalk varies from −30 to −51 dB.

In the ideal, zero aberration, all single-mode demultiplexer, the adjustments being assumed perfect, and in the Gaussian approximation of the single-mode fibre electric field description, the ratio R_w between the width at half-maximum $d\lambda_{Fwhm}$ of the transmission functions and the spectral distance between channels $\Delta\lambda$ is:

$$R_w = \frac{d\lambda_{Fwhm}}{\Delta \lambda} = \frac{1.66 \, \omega_0'}{\Delta x'}$$

($\Delta x'$ and ω_0' being defined as in Section 7.3.1).

This theoretical value is drawn as the continuous curve hereunder (Figure 49). Each cross (+) corresponds to one of the experimental results.

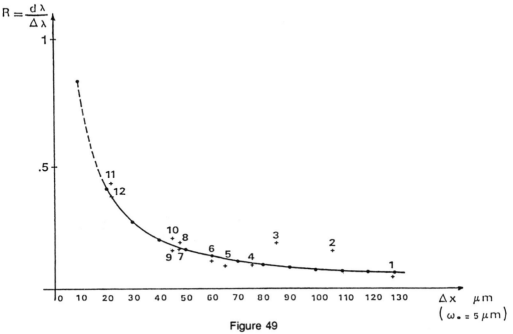

Figure 49
*Relative spectral width at half-maximum
of the all single-mode multi/demultiplexer function of the fibre core distance:*
___ *Theoretical value for perfect multiplexer;*
+ *1–12: experimental points.*

It can be seen that the agreement between theoretical and experimental values is satisfactory, except for experimental points 2 and 3 in which too large aberrations unfortunately increase the spectral width with a corresponding loss.

In order to obtain optimized spectral passbands without requiring a small distance between fibre cores, as necessary on simple devices, different solutions have been proposed such as a spectral recombination within each channel via an intermediate spectrum image through microprisms or through a microlens array, as in [86], in which $R_w = 0.7$ is obtained with up to 32 channels. However, the manufacturing cost and the additive losses of such devices remain a drawback.

8.3 Thermal drift

Thermal drift depends on the thermal expansion coefficients ε of the different materials and on the index variation dn/dt. On monoblock grating WDM, it has been shown [76] that the wavelength shift is given by:

$$\Delta\lambda/\lambda = (\varepsilon + 1/n \; dn/dt) \, \Delta t$$

$\Delta\lambda/\lambda = 0.007$ or 0.005 with silica or $KzFSN_4$ respectively.

> My peace of mind is often troubled by the depressing sense that I have too heavily borrowed from the work of other men.
>
> Albert Einstein

CHAPTER 9

Multiplexers using wavelength selective coupling between fibres

The coherent interaction between optical waveguides has been analysed for many years (for example see [87] – [97]). Placing two fibre cores side by side in interaction over a length L (Figure 50) leads to the basic structure of the x (or y) coupler.

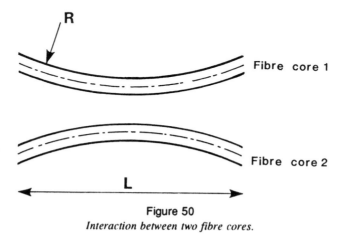

Figure 50
Interaction between two fibre cores.

Theory

Let us consider the case of identical single-mode fibres. It has been shown that the coupling coefficient of the two HE_{11} modes between two parallel fibres, with a distance h between cores, is:

$$c(h) = \frac{\lambda}{2\pi n_1} \frac{u^2}{a^2 V^2} \frac{K_0(\upsilon h/a)}{K_1^2(\upsilon)}$$

a is the fibre core radius, λ is the wavelength, n_1 is the core index, K_0 and K_1 are the modified Bessel functions of second kind of orders 0 and 1, and u and υ are the fibre transverse mode propagation parameters such that $u^2 + \upsilon^2 = V^2$ and V is the normalized frequency at the optical frequency under consideration.

The power coupled from one fibre to the other has the form $\sin^2 C_0 L$ for two parallel fibres and the transmitted power is $\cos^2 C_0 L$; L is the interaction length and C_0 is the coupling coefficient. In fact, as the coupling varies along z, we obtain:

Coupled power: $\sin^2 \int_{-\infty}^{+\infty} c(z) \, dz$

Transmitted power: $\cos^2 \int_{-\infty}^{+\infty} c(z) \, dz$

In order to obtain an overlap of the two fibre modal fields, confined in the cores, and to only a few micrometres, it is necessary to place the cores very close together. This can be achieved by several methods, often by fusing adjacent fibres or by adjustment of fibres after a partial removal of their cladding by polishing. Fused biconical structure couplers using multimode fibres were reported in 1976 [88] to [90]. Polished structure multimode couplers did not appear until 1978, to our knowledge [91]. However, the first papers demonstrating a polished single-mode fibre bidirectional coupler in our bibliography are dated 1980 [92] and [93]. In fused structures, the light transfer is made easier as the fundamental modes spread out when the two cores are reduced by pulling. The cladding can be partially removed by chemical etching [98], but the strength of the fibre becomes a problem unless techniques such as those described by Séverin [99] can be applied. In the polishing technique, each fibre is mounted on a slotted curved glass block (typically, fibre curvature radius = 25 cm). Several approaches are used for the set up in the slotted block and for the polishing [100] (Figure 51). An effective approach is also described in [93].

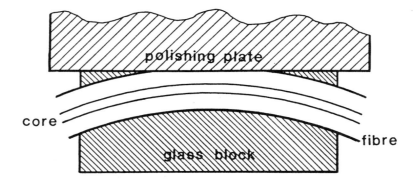

Figure 51
*Directional coupler polishing
(one of the manufacturing methods).*

When two identical fibres are used, the coupling efficiency is a periodic function of the wavelength. The minimum interval between the separated wavelengths $\Delta\lambda$ is given by [94]:

$$\Delta\lambda = \frac{\pi/2}{[\delta(C_0 L)/\delta\lambda]_\lambda}$$

$\delta(C_0 L)/\delta\lambda$ being the derivative of the coupling coefficient (determined by the superposition of the two guides' modal fields) and L being the effective interaction length.

The coupling efficiency is not periodic if the two fibres differ. To a first approximation [95]:

$$\Delta\lambda = \frac{5}{L \left| \dfrac{d\beta_1}{d\lambda} - \dfrac{d\beta_2}{d\lambda} \right|}$$

where $\dfrac{d\beta_1}{d\lambda}$ and $\dfrac{d\beta_2}{d\lambda}$ are the modal propagation constant derivatives of each fibre.

The wavelengths corresponding to an identity of these propagation constants between the two guides are the wavelengths at which an energy transfer takes place from one fibre to the other.

In the fused structure, the taper length is increased until the required coupling value is achieved at the nominated wavelengths. The coupling is measured during elongation. If the fibres are identical, a sinusoidal flux is recorded. The coupler becomes more and more wavelength dependent as the interactive length increases (Figure 52).

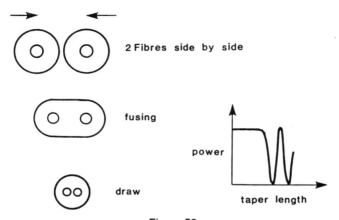

Figure 52
*Fused fibre coupler manufacturing steps
and typical recording of coupled power during elongation.*

The main advantage of the polished single-mode coupler over the fused coupler is its tuneability. The tuning can be scanned by moving the two fibres relative to each other, while the main advantage of the fused coupler wavelength division multiplexer is its manufacturing cost. High performance is obtained by both processes. Insertion losses lower than 1 dB and a power extinction ratio better than 35 dB were obtained more than ten years ago on polished structures [93].

Very small losses (0.04 dB) at wavelengths 1.3 and 1.523 µm were reported by Georgiou and Boucouvalas [101]. A typical wavelength transmission curve of such a device is given in Figure 53. In order to obtain more than two wavelengths, these couplers must be cascaded [103], [104] and [105]. An example is given in Figure 54, with three fused fibre couplers used as a multiplexer at 1320, 1280, 1240 and 1200 nm.

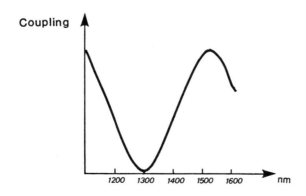

Figure 53
*Typical response on one of the channels of a fused fibre coupler
with two identical fibres. The multiplexer works at 1.3 and 1.5 µm [102].*

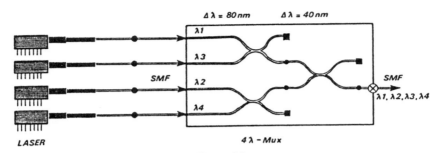

Figure 54
Cascading of fused single mode fibre couplers for multiplexing four wavelengths.
(After Fussgaenger et al.) [103].

The authors obtained from 4 to 6 dB insertion losses and the device is excellent in multiplexing. However, demultiplexer applications require better crosstalk isolation. Consequently, a microoptic grating demultiplexer is usually preferred at the demultiplexer side. The isolation of all-fibre couplers can be improved by passing through several successive couplings. In such a way, Luo et al. obtained 41 dB at 1.3 µm and 50 dB at 1.55 µm of isolation on a two-wavelength 1.3/1.55 µm coupler [106].

L'invention est la faculté de mener à bien une entreprise[1]

Irène Joliot Curie

CHAPTER 10

Trend towards 'integration' of the wavelength division multiplexers and demultiplexers

10.1 Introduction

It would be very desirable to integrate the multiplexer with its optical sources, the demultiplexer with its receivers and the multi/demultiplexer on a single component for a bidirectional link.

As a matter of fact, it is assumed that, as was the case with microelectronics, the integration would facilitate mass production processes, leading to cheap components. Very active research was undertaken on that subject, such as that in [107]. The integration of the structure of the multiplexer or of the demultiplexer, excluding the optical sources and the receivers, was done on planar optical waveguides. But, the full integration with the sources and the detectors is possible on semiconductors only. In such a case, one finds limits due to losses of the materials, losses at the interfaces with the fibres and inherent losses corresponding to the necessary bends in the coupling structures. Losses are also related to the low efficiency of the focusing and dispersive devices used.

Thus, up to now, the results have often been disappointing: exclusively hybrid structures seem operational. Moreover, the costs remain high. The devices may be manufactured from different materials, the best candidates being lithium niobate or lithium tantalate, doped glasses and III-V semiconductors [108].

---

[1] Discovery is the ability to carry a work through successfully.

10.2 Source integration in multiplexing

Among the early results, those of Nakumara *et al* (1976-77) [109]: the integration of frequency multiplexed distributed feedback lasers (DFL) and later of six distributed Bragg reflector lasers (DBR) with 2 nm channel spacing on AsGa; and those of Lee *et al* [110]: the integration of LEDs on AsGa, are significant.

In 1981, an InP LED structure emitting at 1.3 and 1.14 µm from two adjacent areas located within a diameter of 75 µm (Figure 55) was published.

A double GaAlAs laser structure at 849 and 885 nm was published by CNET in 1982 (Figure 56) [111].

Let us also quote Van der Ziel *et al* with an array of five lasers emitting near 1.31 µm and, again, with a spacing of 2 nm [112].

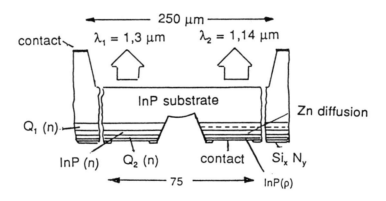

Figure 55
LED 1.3-1.14 µm − BELL, 1981.

Among other examples, there is also the hybrid device by Chin *et al* (Figure 57) [113] with two laser diodes using InGaAsP and InP chips bonded together and that of Koszi and Olsson [114]. In 1987, Dutta *et al* (Figure 58) [115] gave the characteristics of a double laser on InGaAsP in which the two wavelengths were emitted independently and finely tuneable. In 1989, Masashi Nakao manufactured a λ/4 shifted DFB laser array with 20 wavelengths obtained by lithography with a synchrotron beam on an epitaxial InGaAsP structure. The threshold was 15–20 mA and the wavelength domain was about 1530–1550 nm [116].

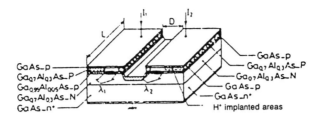

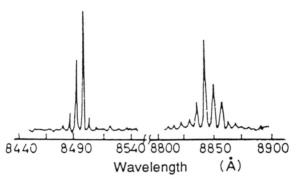

Figure 56
849–885 nm laser – CNET, 1982 [111].

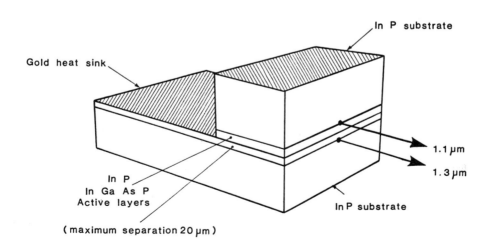

Figure 57
Dual wavelength edge-emitting double diode (Chin et al) [113].

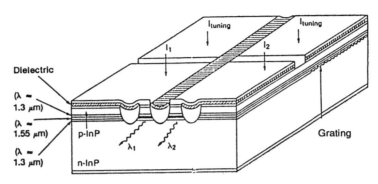

Figure 58
Double laser
(Dutta et al.) [115].

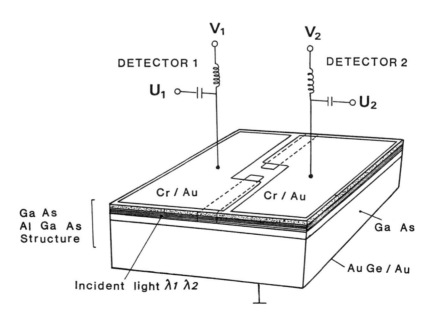

Figure 59
Dual wavelength demultiplexing with integration of the detectors.
(After Larsson et al.) [118].

10.3 Integration of detectors in demultiplexing

The attempts to integrate detectors in demultiplexing began early: 1982 [117]. During the following years, integration of the detectors was actively studied. For instance, in 1986 Larsson and his co-workers achieved a double receiver structure in which the wavelength sensitive range is tuneable with voltages V_1 and V_2 on side by side elements [118] (Figure 59). Another two-wavelength photoreceiver with a resolution of 69 nm around 700-800 nm and 41 dB isolation was obtained in 1989 by Miyazawa et al [119]. In 1990, Ünlü et al made a heterojunction phototransistor using four resonant cavities at different wavelengths between 800 and 940 nm (Figure 60) [120]. This technology was also applied [121] to a wavelength multiplexing optical switch consisting of a resonant cavity-enhanced heterojunction phototransistor (RCEHPT) vertically integrated with a quantum-well, light emitting diode. D. Moss et al have demonstrated a three-and-four-channel WDM detector based on a segmented reverse biased quantum-well GaAs/AlGaAs laser structure. With four channels near 840 nm, the crosstalk was −10 dB, with a wavelength spacing of 14 nm. The response time was less than 40 ps [445].

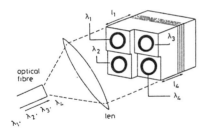

Conceptional application for demultiplexing detector composed of four RCE-HPTs with different surface recessing

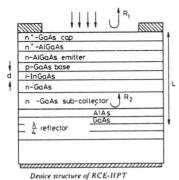

Device structure of RCE-HPT
The resonant cavity is formed between the $\lambda/4$ stack mirror (R_2) below the collector and the semiconductor surface (R_1).

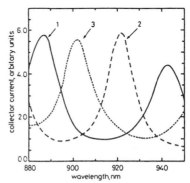

Spectral response of three devices with different surface recessing

Figure 60
Heterojunction phototransistor used in demultiplexing.
From M. S. Ünlü et al [120].

A device integrating a wavelength duplexer for bidirectional communication and a receiver in InP was obtained by Bornholdt et al [122].

10.4 Multi/demultiplexing on planar optical waveguides

10.4.1 Devices using lithium niobate (or lithium tantalate)

Waveguides obtained by titanium diffusion in lithium niobate may be very transparent, for example, at 1.15 µm, 0.09 dB/cm losses were reported [124]; this is comparable to the bulk material loss. The coupling loss between the fibre and such an optical waveguide may be relatively small [125]–[129].

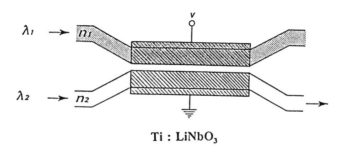

Figure 61
Tuneable integrated optic WDM structure.

An example is given in Figure 61, from [130 - Alferness and Vaselka]. Such a 1.3/1.55 µm active coupler has a 75 nm passband at half-maximum and is tuneable over 120 nm. Intrinsic multiplexing losses of 0.1 dB and a crosstalk of −17 dB were thus obtained. This crosstalk may be reduced by gradual adjustment of the distance between the guides [131]. Using this principle, a four-wavelength multiplexer (749.1, 780.1, 809.7 and 833.9 nm) was achieved in 1989 with an association of three elementary structures (Figure 62) [132]. The crosstalk varies from −18 to −33 dB according to the wavelengths at any port A, B, C or D.

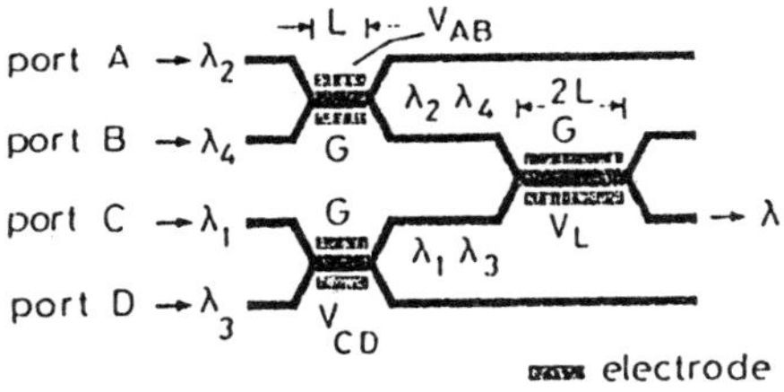

Schematic diagram of four-channel WDM
G = ground

Figure 62
Four-channel WDM on Ti: Li Nb O_3
(From J.P.Lin et al) [132]

In 1989, M. De Sario et al [133] obtained a 1.33-1.55 μm multiplexer with a total insertion loss of 0.9 and 2.3 dB, respectively, and −25 dB crosstalk. Their theoretical analysis starts from an application of the so-called 'Dual effective index method' [134] applied to anisotropic structures.

10.4.2 Devices using amorphous dielectric or semi-conductor waveguides

Planar dielectric waveguides can be used to manufacture passive multi/demultiplexers of different types (coupling mode structure, diffraction grating structure, etc.). Different materials, such as SiO_2, TaO_5, other glasses, organic materials and semi-conductors, were proposed. In the past, small losses have been demonstrated, for instance with GeO_2 by Garside and Jessy (1 dB/cm at 633 nm) [135]. These films were deposited on glass by sputtering.

Other manufacturing methods, including vacuum evaporation, plasma polymerization, ion exchange methods, were proposed (the last being one of the most important and often used in industrial manufacturing).

At the beginning, the proposed structures generally used 'hybrid' planar waveguides with, to some extent, prohibitive manufacturing costs [136] to [139].

Multi/demultiplexers on glass waveguides obtained by ion exchange

Guiding structures can be manufactured from the techniques described in [140], [141] and [142], for instance.

Alkali ion exchange has been used for the last 20 years, and more, for strengthening glass surfaces [143]. This is quite a different application, but the process produces an increase of the surface index that was proposed for the manufacture of integrated optical subtrates (Figure 63). The surface index modification arises out of the combination of two different effects:

- *Density effect:* for instance, when small ions such as Li^+ replace larger ions such as Na^+ or K^+, the glass network collapses around the smaller ions producing a more densely packed structure with, usually, a higher refractive index.

- *Effect related to the electronic polarizability of the exchanged ions:* thus, if ions of larger electronic polarizability such as Tl^+, Cs^+, K^+, etc., replace ions of smaller polarizability such as Na^+, an index increase will result, and vice versa [143]. With Thallium ions [145], 0.1 dB/cm losses were reported as long ago as 1983. An experiment in coupling at 632 and 488 nm (or 514 nm) using the interaction between two joined planar waveguides was described in 1989 by Mohamed Lofti Gomoa and Germain Chartier [146] (Figure 64).

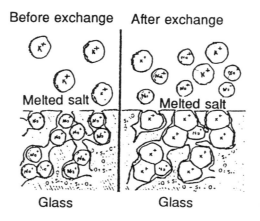

Figure 63
Low temperature ion exchange between K^+ and Na^+.
(Principle after Nordberg *et al*).

In this preliminary experiment, the coupling losses were relatively high (11 dB), but, as was anticipated by the authors, they were greatly improved later on. In France, this technique was developed by Corning Glass [147a,b]. Using the basic principles already reported for fused couplers, achromatic couplers can be manufactured on these waveguides using the standard microelectronic techniques:

photolithography or electron beam lithography, followed by chemical etching, ionic etching, etc. Generally, V-grooves, locations of the input and output fibres, are also etched on the structure. Besides, the structure is buried through a second ion exchange allowing the fabrication of a smaller index interface to optically insulate the guide from the surface.

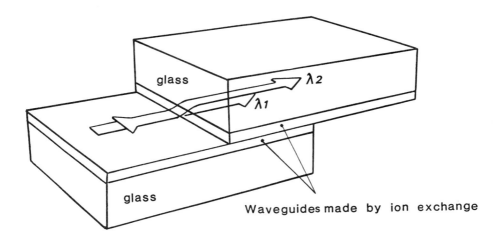

Figure 64
Interaction between planar optical waveguides [146].

One may refer to the patent in [148] as an example of one of the practical manufacturing processes that could be used, or to [149].

Some examples of recent results: in 1989, Goto and Yip, [150] and [151], made a 1.3–1.55 µm multiplexer using an asymmetrical structure. The loading of an Al_2O_3 strip of controlled thickness on the sodalime glass + K^+ guiding structure makes the adjustment of the wavelength range easier to separate (Figure 65). A 10 dB extinction ratio is theoretically obtained over 78 or 45 nm according to the Y angular structure. However, the practical results are slightly worse (5–6 dB).

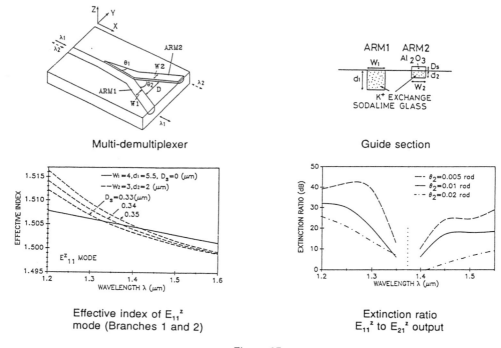

Figure 65
Y branch wavelength multi/demultiplexer for $\lambda = 1.30$ and 1.55 μm obtained on ion-exchanged sodalime glass.
(After N. Goto and G. L. Yip, Spie, Vol. 1141, ECIO'89, 5th European Conference on Integrated Optics.)

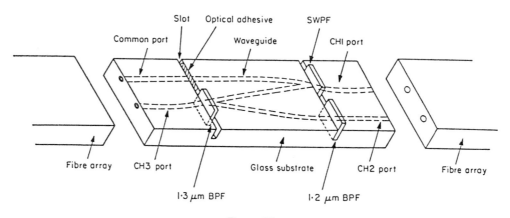

Figure 66
Schematic view of the guided-wave multi/demultiplexer.
(After M. Seki, R. Sugawara and Y. Hanada [152].)

In 1989, Seki et al [152] published the device shown in Figure 66, using a borosilicate-type glass waveguide obtained by $Na^+K^+ \Leftrightarrow Tl^+Cs^+$ ion exchange. On this component, multidielectric, multicavity filters allowing the separation of 1.2 and 1.3 µm wavelengths with a good spectral width (66 nm) are added in etched slots. The losses are 1 dB for the demultiplexer and 2 dB for the multiplexer (the waveguides are multimode). The same year, Negami et al [153] announced the results of an asymmetrical Y junction separating 0.63 and 0.84 µm. This structure is manufactured on a Corning 7059 glass with a double ion exchange. It is a monomode structure.

In 1990, Suzuki et al released excellent results from a 1.3/1.5 µm single-mode multiplexer. The main characteristics of this device can be seen in Figure 67 [154]. In addition, in 1989, Barlier, Nissim and Dohan [155] published 20 dB isolation over 25 nm and 4 dB losses from a 1.31/1.55 µm Y component. At the same wavelengths, using a Mach-Zehnder interferometer structure, −30 dB crosstalk was obtained by Tervonen et al [156]. The ion exchange waveguides can also be used in diffraction grating devices. In this way, N. Kuzuta and Hasegawa manufactured a demultiplexer on a thallium ion exchange multimode waveguide. The thickness was about 70 µm. A flexible concave replica grating was bonded on the convex edge of the waveguide. This device uses 1200, 1310 and 1550 nm wavelengths with 2 to 2.4 dB losses [157] (Figure 68).

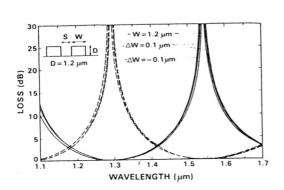
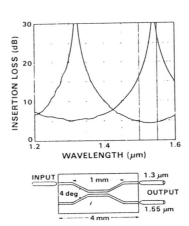

Figure 67
Single-mode multiplexer of S. Susuki et al.
Transactions of the IEICE, vol. E73, n°1, January 1990 [154].

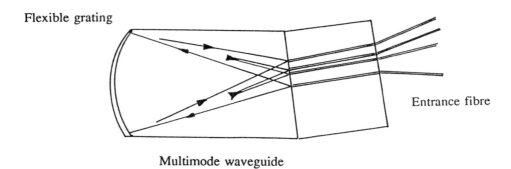

Multimode waveguide

Figure 68
Multimode waveguide grating device.
Noboyuki Kuzuta and Eiichi Hasegawa, *J.Appl.Phys.* (Oct.88) [157].

Semi-conductor devices

Silicon substrates

An optical demultiplexer using a Corning 7059 glass waveguide with an optical isolation from the silicon substrate by a SiO_2 coating was proposed by Masayuki Takami (Figure 69) [158].

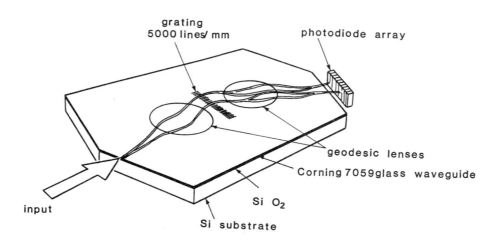

Figure 69
Integrated optical demultiplexer.
(After Masayuki Takami.)

Multi/demultiplexing on planar optical waveguides 85

The focusing is obtained through the use of geodesic lenses, and the dispersion through an ion etched 5000 lines/mm grating, but the losses remain relatively large: 10 dB including the fibre coupling losses. The multiplexer from Leti Laboratories [159] (Figures 70a, b, c) uses a silicon planar waveguide on silicon. The wavelength separation is obtained with a concave grating etched on the waveguide. The entrance and exit fibres are located in V grooves etched into the silicon, for a four-channel device with 20 nm between 7.41 and 8.16 dB. A larger number of channels has also been obtained [160].

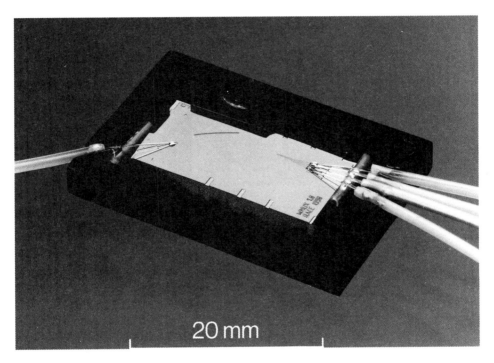

Figure 70a
Multiplexer from Leti (Leti phototech.) [159].

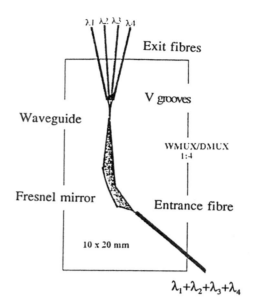

Figure 70b
Leti Laboratories multiplexer, schematic view [159], [160].

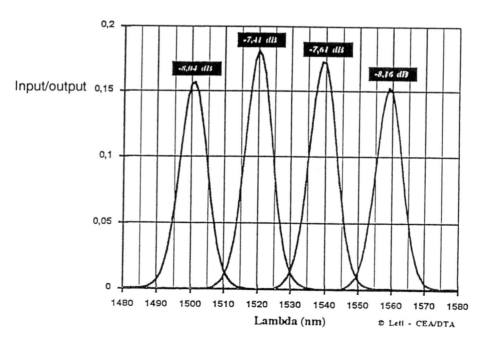

Figure 70c
Leti multiplexer spectral transmission.

A schematic view of another silicon structure made in 1990, using a SiO_2 substrate, can be seen in Figure 71 [161]. A concave grating is used in this device.

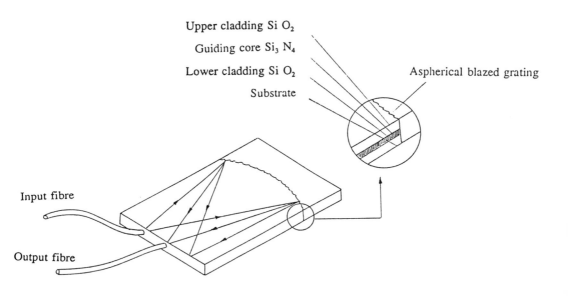

Figure 71
Spectrometer with grating integrated on a silicon substrate.
(After Nasa Tech. Briefs, April 1990.)

In the Bell Labs., a relatively simple planar processing technique, the so-called 'Hybrid optical packaging on silicon' was used to build a four-channel multiplexer with an array of four Bragg reflectors with 3 dB insertion loss [162].

A fibre-to-fibre insertion loss of 10.1 dB and a crosstalk attenuation larger than 15 dB have been achieved with an aberration-corrected, concave grating, flat field spectrograph in silica glass on a silicon device, with 4 nm spacing near the 1.54 µm wavelength [446].

InP substrate

The large losses formerly obtained with these semiconductor waveguides can be reduced by the absorption decrease of free carriers by using a high-purity substrate and weakly doped GaAs guiding layers [163]. As early as 1985, losses smaller than 2.5 dB cm^{-1} on heterostructures were obtained by molecular beam epitaxy [164]. De Bernardi *et al* [165], using an InP substrate on which AlGaInAs quaternary layers were epitaxied, succeeded in manufacturing waveguides with losses smaller than 1 dB/cm^{-1}. X couplers, made by using this waveguide, demonstrated multiplexing with 35 to 100 nm spectral range between channels and an isolation better than 17 dB (Figure 72).

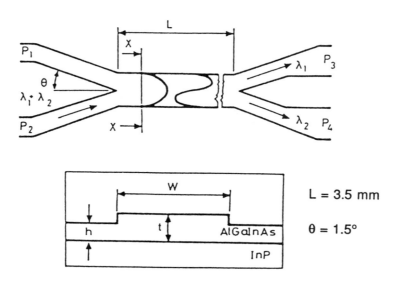

Figure 72
Wavelength division demultiplexer on Al Ga In As/In P for 1.5 µm
(After C. De Bernardi, Elect. Lett., vol.25, n°22, 26 Oct 89.)

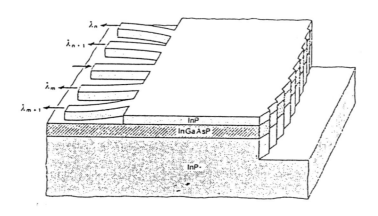

Figure 73
Cremer et al concave grating device [168].

Using this basic technology and a triple-moded rib waveguide '3 mi', other multiplexers which were more dispersive, with 30 nm channel spacing could be obtained [166]. A 1.55–1.3 µm Y multiplexer InGaAsP on InP is given in [167].

Concave grating devices made with guides of the same type were proposed. C. Cremer's device (Figure 73) [168] can work from 1.3 to 1.6 µm. It is compact: 1 x 3 mm². The grating is obtained by holographic methods.

Likewise, the J. B. D. Soole device was manufactured using an etched concave grating. It can separate 50 channels with 1 nm spacing, 0.3 nm channel width and a 19 dB isolation between channels [169]. The manufacturing of narrowband add-drop wavelength multiplexers integrated with DBR lasers has been demonstrated [170].

Remarks

• In the Bell Labs., using the 'photonic integrated circuit' technology, a three-wavelength WDM transmitter, with three MQW-DBR tuneable lasers, a 3 to 1 coupler and an optical amplifier, was built on a semiconductor chip [171a]. Another MQW–DFB laser array module with four channels is described in [171b].

• Holographic phase gratings can be recorded on polymer films coated on different materials [172]. Using this technique, Wang *et al* made wavelength demultiplexers on sodalime glass. This is interesting but a better grating coupling efficiency is still required. Demultiplexer outside coupling using the focusing of a specific holographic recording was demonstrated by Suhara *et al* [173] (Figure 74) and others [174]–[176]. In this specific recording, interferences at a given wavelength between waves coming from a point outside the guide and waves inside the guide are used to record a grating on the guide. Vice versa in demultiplexing, the waves inside the guide are coupled out by the recorded grating with a wavelength dispersion.

• A grating demultiplexer or filter can be produced directly on the fibre itself [177] and [178] (Figure 75).

• In order to increase resolving power, Takahashi *et al*, [179] and [180], proposed to increase the optical path difference between 'diffracting' elements using the waveguide structure shown in Figure 76.

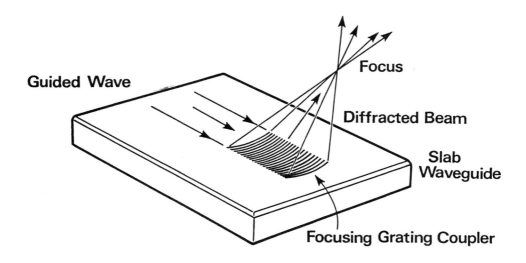

Figure 74
Holographic demultiplexer coupler.

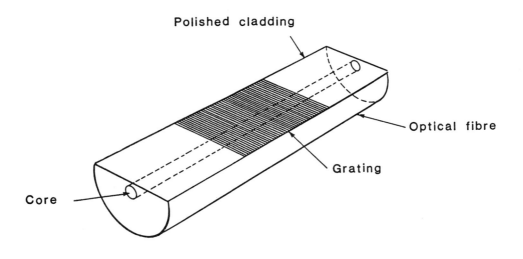

Figure 75
WDM using a grating etched directly on the fibre core.

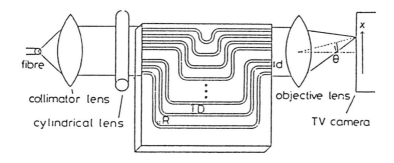

Figure 76
Arrayed waveguide grating with an increase of optical path difference between 'diffracting' elements for nanometric resolution **(Takahashi et al [179])**.
Length difference between adjacent channels: $\Delta L = 2(D-d)$.
Transmission for wavelengths such that $n_c \Delta L + n_s d \sin\theta = m\lambda$.

> *It should be noted that in information theory, organization is equated with information.*
>
> Sir Julian Huxley

CHAPTER 11

Some techniques used in wavelength division multiplexing

11.1 Light Emitting Diode (LED) spectral slicing

11.1.1 Introduction

The use of wavelength division multiplexing generally provides flexibility in the elaboration of solutions and the evolution of networks. But the cost saving related to reducing the number of fibres (this may result in further civil engineering, cable and connection cost savings) is sometimes counterbalanced by extra costs resulting from more stringent specifications of optical sources. If the bitrate of each elementary signal is not too large and if the transmission lengths are short enough, it is nevertheless possible to use low cost light emitting diodes (LED) (up to some tens Mb/s and a few kilometres; typical applications: process control networks, telephone links, etc.). The choice of a LED with well-separated wavelengths is limited, but the LED spectral slicing technique, that we will discuss now, allows a large increase in the number of channels. As early as 1982 [37], we pointed out that for short-distance links and low bit rates, one could tolerate a specified spectral overlap or even multiplex signals from identical diodes, and in 1983 [38], we indicated that the multiplexing of twenty identical LEDs could be obtained with −45 dBm signals at the receivers for 2 km and 50/125 µm fibre links with the LEDs available at that time. These results were obtained from our experiments on an optical sensor network. Using this technique, a set of LEDs, identical or with adequate spectral shifts, is typically connected to the multiplexer entrances. The multiplexer cuts well-separated spectral slices, one different slice for each entrance. The specifications of a 42-channel multiplexer obtained with

LED spectral slicing

1 nm slices widths, cut in a LED emitting around 820 nm and used in a process control network with 100/140 µm fibres, were given in 1984 [181]. At the present time, single-mode fibres are more often used, even in spectral slicing: M. H. Reeve *et al.* [182] described a four-channel network with 20 nm spacing between channels around 1300 nm and at 2 Mb/s. Use of this technique in the UK loop network has been reported by Hunwicks *et al.* [183]. The use of superluminescent diodes allowed the extension of the spectral slicing up to sixteen channels as early as 1990 [184].

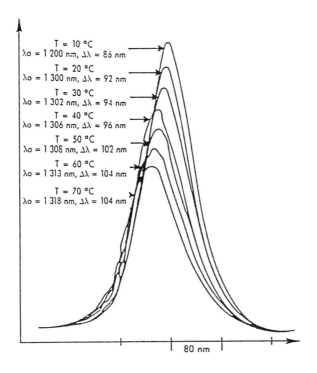

Figure 77
*Typical emission spectrum at different temperatures of a commercial 1300 nm LED.
(CIT Alcatel.)*

The effect of source temperature variation (an example is given in Figure 77) was analysed in detail by Hunwicks [185] at 1300 nm. Considering a 0.54 nm/°C thermal shift from 0 to 70 °C, he showed that additional losses appear of up to 3.8 dB for a four-channel multiplexed system, and 4.7 dB for ten channels (single-mode fibre case). Bersiner and Rund [186] demonstrated bidirectional transmission in five channels over 8 km of standard GI fibres at bit rates of 167 Mbit/s per channel in the 1.3 µm domain using InGaAs avalanche photodiodes.

The characteristics of multiplexers for LED slicing are given in [39] (Figure 78): for example, with ten single-mode fibre channels, spectral spacing between channels of 9 nm at 1300 nm and FWHM of the wavelength transmission functions of 1.6 nm on the multiplexer and 6 nm on the demultiplexer, the combined losses for the multiplexer and the demultiplexer are under 5 dB, and the optical crosstalk is lower than −31 dB.

The use of spectral slicing for wavelength-routed subscriber loops, with superluminescent diodes at 150 Mbit/s over ten channels or 50 Mbit/s over sixteen channels and 7 km in a first demonstration, and with LEDs over four channels, without optical amplification, was experimented by T. E. Chapuran *et al* [447]. In a third experiment, these authors showed how erbium-doped fibre amplifiers can greatly alleviate the power-budget constraints of spectral slicing, as it is possible to keep the same performance in networks connecting splitter-based fibre loop architectures to a broadband switching hub. These experiments show how the amplifiers can mitigate the effects of splitting losses as well as other sources of attenuation. G. J. Lampard [448] arrived at an equivalent conclusion: in networks using spectral slicing and optical amplification, the capacity can be increased to 30 channels operating at 60 Mbit/s.

11.1.2 Theoretical analysis

All single-mode fibre case

Let us assume that the transmission fibre is single-mode along with all the entrance fibres of the multiplexer. We have seen that the multiplexer spectral transmission function is Gaussian if the system has no aberrations. We assume, in addition, that the diode spectrum can be approximated by a Gaussian function (Figure 79). These functions have the following expressions:

- for the multiplexer: $F' = \exp\left[-4\ln 2 \left(\dfrac{\lambda - \lambda_m}{\Delta\lambda_{mux}}\right)^2\right]$

- for the emitter: $G' = \exp\left[-4\ln 2 \left(\dfrac{\lambda - \lambda_e}{\Delta\lambda_{em}}\right)^2\right]$

In these equations, we introduced the full widths at half-maximum (FWHM), $\Delta\lambda_{mux}$ for the multiplexer and $\Delta\lambda_{em}$ for the emitter. λ_m is the wavelength at a specific passband maximum of the multiplexer and λ_e is the wavelength at the emitter maximum.

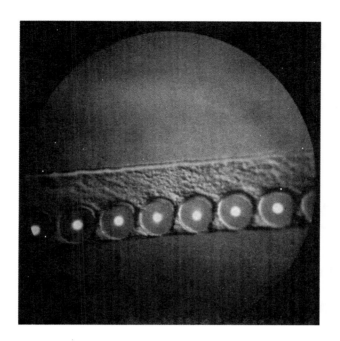

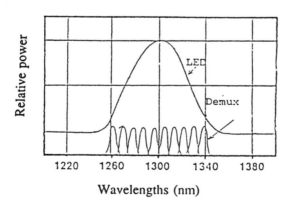

Figure 78
*Above: partial cross-section of the single-mode fibre array;
Mode diameter 10.8 μm, distance between fibres 47.5 μm.
Below: characteristics of the emitter and passbands in the slices.*
(J.-P. Laude *et al*) [39].

The transmission is given by the convolution product: $[F' * G']$.

To simplify the expression, we use the variables:

$$x = \lambda_e - \lambda_m, \quad \omega_0 = \frac{\Delta\lambda_{mux}}{2\sqrt{\ln 2}} \quad \text{and} \quad \omega_1 = \frac{\Delta\lambda_{em}}{2\sqrt{\ln 2}}$$

$f(x) = \exp -\pi x^2$ being a Gaussian function, it is known that its Fourier transform is $\bar{f}(u) = \exp -\pi u^2$

We have already seen that the function $f'(x) = f(x/a)$ has a Fourier transform $\bar{g}'(u) = |a|\,\bar{f}'(au)$.

Therefore, the Fourier transform of $f'(x) = \exp -\pi(x/a)^2$ is $\bar{f}(u) = |a|\exp -\pi a^2 u^2$. Similarly, the Fourier transforms of $f_0(x) = \exp -(x/\omega_0)^2$ and of $g_0(x) = \exp -(x/\omega_1)^2$ are respectively $\bar{f}_0(u) = \sqrt{\pi}\,|\omega_0|\exp -\pi^2\omega_0^2 u^2$ and $\bar{g}_0(u) = \sqrt{\pi}\,|\omega_1|\exp -\pi^2\omega_1^2 u^2$.

If:

$$F(x) = \exp -(x/\omega_0)^2$$

$$G(x) = \exp -(x/\omega_1)^2$$

$$[F * G] = \left[\exp -(x/\omega_0)^2 * \exp -(x/\omega_1)^2\right]$$

Then:

$$[F' * G']_{\lambda = 0} = [F * G]_{x = \lambda_e - \lambda_m}$$

LED spectral slicing

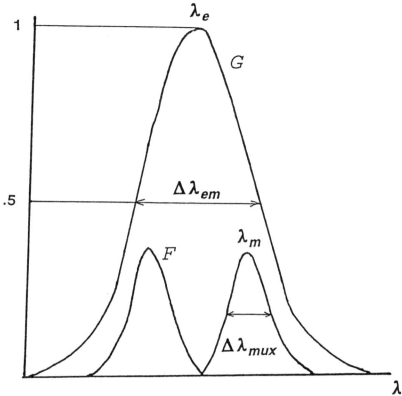

Figure 79
Theoretical losses in LED slicing.
G: LED emission curve;
F: multiplexer channel transmission.

The transform of [F ∗ G] is the product of the Fourier transforms of F and G respectively:

$$\text{TF}\,[F * G] = \pi\,|\omega_0|\,|\omega_1|\,\exp -\pi^2(\omega_0^2 + \omega_1^2)u^2$$

so:

$$\frac{\pi\,|\omega_0|\,|\omega_1|}{\sqrt{\pi\,(\omega_0^2 + \omega_1^2)}}\,\sqrt{\pi(\omega_0^2 + \omega_1^2)}\,\exp -\pi\left(\sqrt{\pi\,(\omega_0^2 + \omega_1^2)}\,u\right)^2$$

This last expression has the form $|a|\,\exp -\pi a^2 u^2$

Then, the inverse Fourier transform is:

$$[F * G] = \frac{\sqrt{\pi}\,|\omega_0||\omega_1|}{\sqrt{\omega_0^2 + \omega_1^2}} \exp\left(-\left(\frac{x}{\sqrt{\omega_0^2 + \omega_1^2}}\right)^2\right)$$

If the FWHM ω_0 of the multiplexer is large compared to the FWHM ω_1 of the emitter, and if the functions are centred ($\lambda_e = \lambda_m$):

$$[F * G]_0 \rightarrow \sqrt{\pi}\,|\omega_1|$$

This corresponds to the case where the LED is not sliced and is the maximum transmission value achievable.

The multiplexer transmission factor will be:

$$\frac{[F * G]}{[F * G]_0} = T$$

$$T = \frac{1}{\sqrt{1 + \left(\frac{\omega_1}{\omega_0}\right)^2}} \exp\left(-\left(\frac{x}{\sqrt{\omega_0^2 + \omega_1^2}}\right)^2\right)$$

We return to the physical notations. Let us write:

$$x = \lambda_e - \lambda_m$$
$$\omega_0 = \frac{\Delta\lambda_{mux}}{2\sqrt{\ln 2}}$$
$$\omega_1 = \frac{\Delta\lambda_{em}}{2\sqrt{\ln 2}}$$

This becomes:

$$\boxed{T = \frac{1}{\sqrt{1 + \left(\frac{\Delta\lambda_{em}}{\Delta\lambda_{mux}}\right)^2}} \exp\left(-4\ln 2\, \frac{(\lambda_m - \lambda_e)^2}{\Delta\lambda^2_{mux} + \Delta\lambda^2_{em}}\right)}$$

Demultiplexer contribution (Figure 81)

1. If the demultiplexer is spectrally adjusted with all single-mode fibres and is identical to the multiplexer, we will obtain a demultiplexer additional loss related to the cross correlation of the Gaussian functions of the multiplexer and of the demultiplexer; these functions are identical:

$$T_{DM} = \frac{1}{\sqrt{2}} = 0.7071$$

$$T_{TOTAL} = \frac{1}{\sqrt{2\left(1 + \left(\frac{\Delta\lambda_{em}}{\Delta\lambda_{mux}}\right)^2\right)}} \exp - 4\ln 2 \frac{(\lambda_m - \lambda_e)^2}{\Delta\lambda^2_{mux} + \Delta\lambda^2_{em}}$$

2. If the demultiplexer uses multimode exit fibres, the transmission functions are larger and the theoretical loss contribution of the demultiplexer is negligible. If this component has no absorption and wavelength channels adjusted to those of the multiplexer, then:

$$T_{TOTAL} = \frac{1}{\sqrt{1 + \left(\frac{\Delta\lambda_{em}}{\Delta\lambda_{mux}}\right)^2}} \exp - 4\ln 2 \frac{(\lambda_m - \lambda_e)^2}{\Delta\lambda^2_{mux} + \Delta\lambda^2_{em}}$$

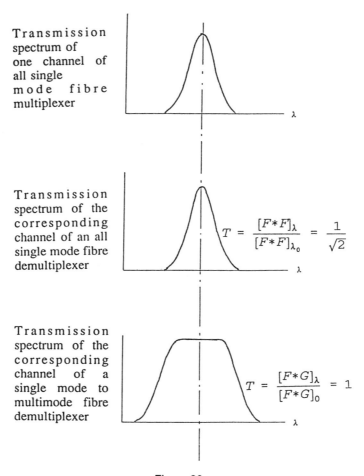

Figure 80
Theoretical demultiplexer losses, T.

11.1.3 Calculation of the spectral filtering losses of real systems

First example corresponding to the actual case of Figure 81.

This is an all single-mode fibre multiplexer with the following characteristic ratio:

$$\frac{\Delta \lambda_{mux}}{\text{Channel spectral distance}} = 0.44$$

LED spectral slicing

To date, using classical multiplexers, this ratio of 0.44 is the best that can be achieved with small crosstalk with an all single-mode multiplexer [84].

The emitter emission curve is approximated to a Gaussian function with FWHM of: $\Delta\lambda_{em} = 61$ nm; the multiplexer has a Gaussian transmission function with $\Delta\lambda_{mux} = 22$ nm.

We take $\lambda_m - \lambda_e = 25$ nm:

$$T = 0.225$$

Thus, the multiplexer loss due to slicing is 6.5 dB.

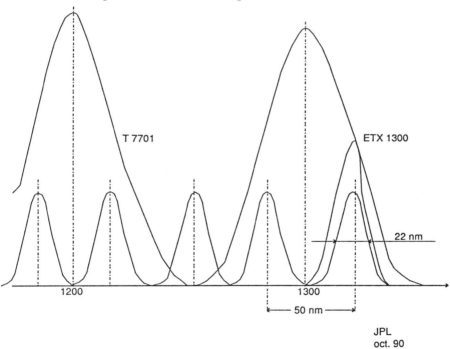

JPL
oct. 90

Figure 81
Typical spectral slicing multiplexing: two LEDs, four channels.
Upper curves: emission spectra.
Lower curves: multiplexer transmission channels.
On right inside: product of emission x transmission.

Second application example

Let us consider M. H. Reeve *et al.*'s device [182]. This is a single-mode, four-channel multiplexer at 1270, 1290, 1310 and 1330 nm, $\Delta\lambda_{mux}$ = 3.65 nm, $\Delta\lambda_{em}$ = 63 nm. For the worst channel $\lambda_e - \lambda_m$ = 30 nm, and T = 0.0218 for the multiplexer + demultiplexer, that is, a theoretical loss of − 16.6 dB. Thus, for − 25 dBm injected power, 41.6 dBm remain after slicing losses due to the multiplexer and the demultiplexer. Noting that the losses due to component defects must be added to this, our theoretical calculation is in agreement with the values given in [182].

Third application example

Let us assume that the devices are those of [39]. The multiplexer is an all single-mode fibre, ten-channel component. On the demultiplexer, multimode 50/125 μm fibres are used, so the 'shaping' losses due to the demultiplexer can be neglected. The transmission function characteristics of the multiplexer and demultiplexer are given in Figure 82 and Figure 83. If an emitter centred on 1300 nm and with $\Delta\lambda_{em}$ = 90 nm is used, we find that T = 0.0100 to 0.0171 from the channel near the emitter maximum to the edge channels; this corresponds to + 20 to + 17.7 dB losses with such an emitter; 10 to 20 μW (− 20 dBm to − 17 dBm) can be injected. Consequently, we obtain − 40 to − 37 dBm in the worst channel. With a receiver of − 56.5 dBm (10^{-9} BER) sensitivity at 2 Mb/s, a 16.5 to 19.5 dB margin remains available for the transmission line, including 4 to 6 dB of multiplexer + demultiplexer typical loss.

Note: In 1992, an LED specially designed for spectral slicing became available (830 nmn 120 nm FWHM, 400 μW coupled in 400 μm core).

LED spectral slicing

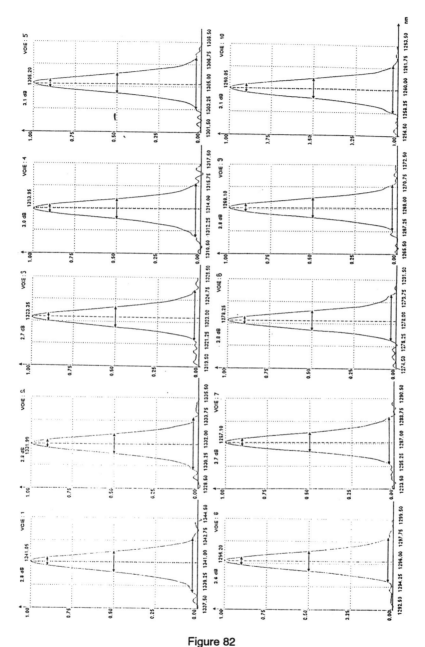

Figure 82
*Transmission functions of a ten-channel all SM fibre multiplexer.
Distance between channels $\Delta\lambda = 9$ nm, $\Delta x = 47.5$ μm, FWHM = 1.6 nm.*

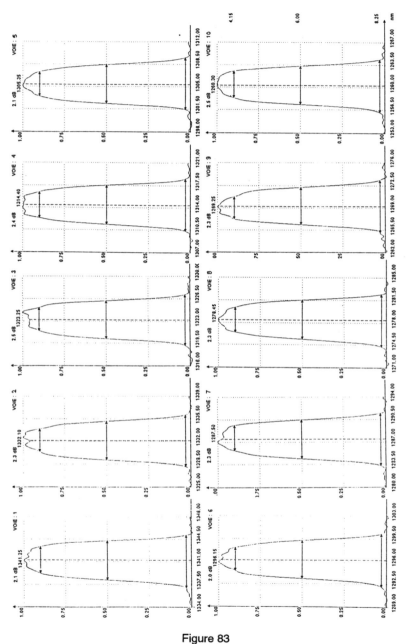

Figure 83

Transmission functions.
Ten-channel demultiplexer with SM fibre in and 50 x 125 GI fibres out.
Separation between channels Δλ = 9 nm, FWHM = 6 nm.

11.2 Shared optical function technology wavelength division multiplexing and coupling ('SOFT WDM' and 'SOFT couplers')

11.2.1 General principle

A coupling function or a wavelength multiplexing function through an optical system with two or three dimensions generally sets a relation between a transmission fibre and an entrance or exit fibre set. Thus, for instance, in a Y-coupler, the transmission line fibre is imaged on to two other fibres through an optical divider (semi-transparent mirror or pupil divider) and one or more focusing optics. In a wavelength division multiplexer, the transmission fibre is imaged according to the wavelength at different focal plane locations with a grating (and/or a set of multidielectric filters), these locations being associated with one or several focusing optics (except in the concave grating case which combines dispersion and focusing functions). As a matter of fact, all these devices provide a relation between object and image fields (in so far as the stigmatism is sufficient and the relative fibre positions in the different fields are controlled, one could realize several couplers and/or several elementary multiplexers, through the same dividing and focusing optics). We call these devices 'shared optical function technology' ('SOFT') components [187].

In a 'SOFT' component, one coupling optic is used on a set of P fibres arranged in subsets of n+1 fibres. So, for instance, from a p = 21 fibres array, the same optics set the coupling or the duplexing of one fibre to n = 2 fibres, p/n + 1 = 7 times. (We obtain seven identical couplers or duplexers in only one component (Figure 84)). This principle will be particularly attractive for cost saving networks in which several couplers or multiplexers are located in the same area, for example, in a central office, a remote terminal or a branching point of a videocommunication network.

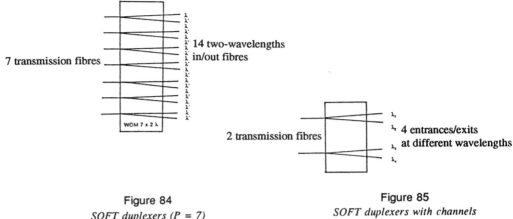

Figure 84
SOFT duplexers (P = 7)
at two identical wavelengths.

Figure 85
SOFT duplexers with channels
at different wavelengths.

On a fixed device, each set of different entrance and exit fibres can correspond to a collection of different wavelengths, i.e. the multiple multiplexers need not necessarily be identical. This is particularly interesting for the design of fibre spectrometers with many channels or for telecommunication demultiplexers, as well as for the design of versatile multiplexers in which wavelength channels can be adapted to the spectral range of the LED sources' spectral range development or can be adapted to the evolving availability of optical telecommunication lasers.

11.2.2 The particular case of grating multiple multiplexers

Let us examine the grating multiple multiplexer case. We first assume a single array of fibres. See another schematic view in Figure 86. A set of p fibres is located in the focal plane F of an optical system composed of a focusing lens O and a grating R, the fibres are numbered from 1 to p, i and j are the references of any fibre pair. If λ_{ij}, the wavelength in the medium in front of the grating, corresponds to the coupling between the entrance fibre i and the exit fibre j in the first order of the grating, we obtain:

$$a (\sin \alpha_i + \sin \alpha_j) = \lambda_{ij}$$

If the position of the fibres in the focal plane is such that: $\sin \alpha_{i+1} = \sin \alpha_i + u$, where u is a constant, then we also obtain:

$$\sin \alpha_j = \sin \alpha_i + (j - i) u$$

and

$$a [2 \sin \alpha_i + (j - i) u] = \lambda_{ij}$$

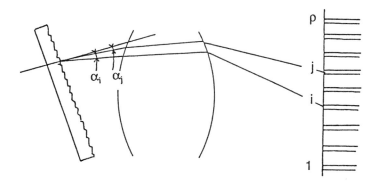

Figure 86
*Schematic view of a grating multiplexer R
with its focusing lens and the fibre plane.
The rays coming from each fibre are parallel in front of the grating.*

It is easily demonstrated that $\lambda_{ij} = \lambda_{ji}$ and that $\lambda_{ij} = \lambda_{i+p, j-p}$. Thus, the matrix corresponding to λ_{ij} of Figure 87 is symmetrical relative to its $i = j$ diagonal and the wavelengths corresponding to the straight lines such as $i + j$ = constant are equal. Note that the wavelengths such as λ_{i+j} are given by an arithmetical progression of a·u difference.

Application to n wavelength multiple multiplexer design for optical telecommunication

With a set of p fibres, one can theoretically multiplex or demultiplex n wavelengths $p / (n + 1)$ times; p and n are necessarily such that $p / (n + 1)$ is an integer. This can be done without associated additive losses and the elementary multi/demultiplexers are independent (except for crosstalk). It is possible to choose the entrance and exit fibre locations from the matrix. One can proceed as follows:

1• Enter through fibre 1, exit through fibres p, p − 1, p − 2, p + 1 − n, (p being the maximum number of fibres and n the number of wavelengths to be multiplexed or demultiplexed).
2• Draw the straight lines $i + j$ = constant such as $i + j = p + 1$, $i + j = p$, $i + j = p − 1$,..., $i + j = p + 2 − n$.
3• Enter through fibre p − n, the intersection of the horizontal line p − n with the previous $i + j$ = constant lines gives the wavelengths corresponding to each exit fibre.
4• Enter through fibre p − 2n − 1 and repeat the same operation
5• etc.

108   *Some techniques*

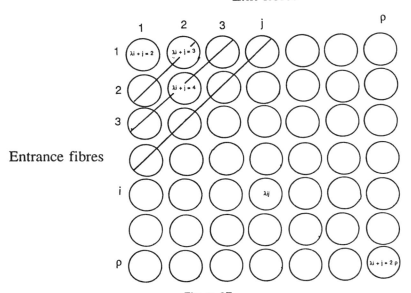

Figure 87
Matrix giving entrance to exit fibre wavelength relation. Note the symmetry relative to the diagonal $i = j = 1$ to $i = j = p$ and the fact that the wavelength is unchanged along straight lines perpendicular to the diagonal. λ_{ij} linearly increases with $i + j$.

(a)   Three times two wavelengths multiplexer

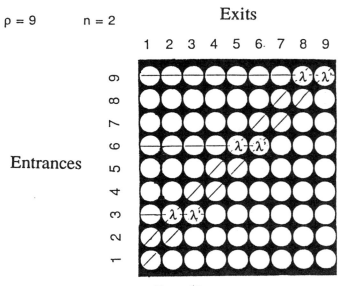

Figure 88
Wavelength fibre matrix on a triple duplexer (nine fibres in the focal plane $p=9$, $n=2$).

The relation between entrance fibres, exit fibres and wavelengths is given in the following diagram (Figure 89):

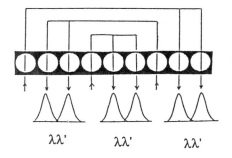

Figures array at the focal plane

Figure 89
Relation between entrances, exits and wavelengths of a triple duplexer. The spectral transmissions are given here for a grating component with 80 x 125 GI adjacent fibres.

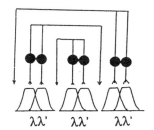

Figure 90
Relation between entrances, exits and wavelengths of a triple duplexer. The spectral transmissions are given here for a grating component with entrances on 30/50 GI adjacent fibres and an exit on 80/125 GI fibre.

(b) Two times three-wavelength multiplexer

$$p = 8 \qquad n = 3$$

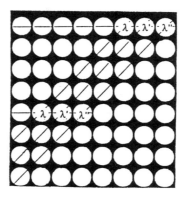

Figure 91
Wavelength/fibre matrix on a double three-wavelength multiplexer (eight fibres in the focal plane) $p = 8$, $n = 3$.

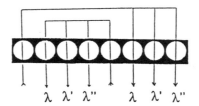

Figure 92
Relation between entrance fibres, exit fibres and wavelengths in the focal fibres plane of a double three wavelengths multiplexer.

As a general rule, for p fibres, the matrix contains 2p−1 different wavelengths in a given grating order, if it is assumed that the same fibre can be used for the entrance and exit, and k(2p−1) wavelengths if k grating orders are used.

11.2.3 Remarks

In Section 11.2.2, we have considered all fibres (or emitters or receivers) along a straight line as this will necessarily be the case in integrated optics. However, one can make use of the two dimensions of the focal plane in microoptics three dimensional devices. For instance, one can place several fibre rows one above the other. Most frequently, two fibre rows are used for double multiplexer manufacturing. Another application of the 'SOFT WDM' can be seen in Chapter 14, in a component with an array of 50 entrance fibres parallel to the grating grooves and 50 spectra imaged on a CCD detector. The device is equivalent to 50 demultiplexers each having several hundred wavelengths.

11.3 Tuneable demultiplexers

In some network architectures, each customer obtains several high bit rate services on different closely spaced wavelengths. The different services are broadcast together on the network and the customer selects one of the services using a tuneable demultiplexer. Such a demultiplexer may use a diffraction grating with piezoelectric tuning (Figure 93a) or, more often, Fabry-Perot or acoustooptic deflector devices. Other devices such as Mach-Zehnder interferometers [455], distributed feedback wavelength filters [456] and selective optical amplificators, are also used.

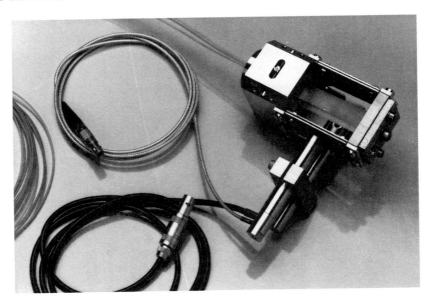

Figure 93a
Jobin-Yvon piezoelectric tuneable grating demultiplexer
(FWHM: 3 nm, range: 1500-1560 nm, loss: 3dB, drift: 0.05 nm/°C).

11.3.1 Fabry-Perot devices

In an elementary Fabry-Perot device [188]-[192] with two plane and parallel plates, thickness t and index n between plates, used with a parallel beam at an external angle i with the perpendicular to the plates, it can be shown that the transmitted intensity in the direction i is such that:

$$I(\sigma) = \left(\frac{T}{1-R}\right)^2 \frac{1}{1 + \frac{(4R \sin^2 \pi\sigma\Delta)}{(1-R)^2}}$$

R being the reflecting power of a plate and T its transmission.
$\sigma = 1/\lambda$ is the wavenumber and $\Delta_0 = 2$ nt cos i is the optical path difference between two successive rays.

Thus, one obtains a set of transmission peaks with maxima $(T/(1-R))^2$ if the materials have neither absorption nor diffusion. The spectral selection can be obtained through the variation of n, t or i. Wavelength selection is easily obtained even if wavelengths are closely spaced. The use of two or more Fabry-Perot (FP) filters, optically in series, allows free spectral range between transmission peaks to be increased at a given resolution. Eng et al [193] designed a two-stage tuneable FP in such a way, with a free spectral range of 15 000 GHz and a finesse of 5170 to be used as a potential 1000 x 1000 optical switch. The miniaturization of the FP device or its integration on an optical chip is feasible. C. M. Miller [194] gives the characteristics of a FP used in a long-haul experiment and in the IBM Rainbow WDMA network: with losses lower than 2 dB and finesse larger than 100.

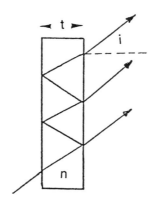

Figure 93b
Fabry-Perot device - Notations.

In spectrometry, Fabry-Perot filters, in which the cavity length t can be tuned piezoelectrically, have been used for more than twenty years and have been

Tuneable demultiplexers 113

adapted to the new telecommunication needs. For example, in [449], such filters are described with finesse values in the range 100-500 and with insertion losses of between 3 and 8 dB depending on the finesse.

The index n can be controlled in liquid crystal Fabry-Perot devices by low driving voltages [450]. For example, staking two liquid crystal FP interferometers, finesse values of 1800, and resolution down to 0.2 nm, were obained [451]. Temperature tuning was also reported [452].

11.3.2 Tuneable acoustooptic filters

The tuneable acoustooptic filter was proposed by Harris and Wallace in 1969 [201].

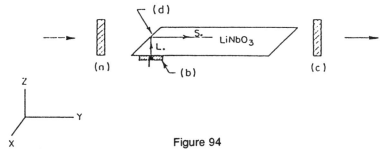

Figure 94
Harris and Wallace acoustooptic filter [201].

In the initial experiment, a $LiNbO_3$ crystal was set between a crossed polarizer and analyser. The incident electric field after transmittance through the polarizer is parallel to the crystal optical axis z and propagates at extraordinary index. Without the acoustic wave, there is an extinction through the analyser at the crystal exit. When an acoustic wave is applied in A and directed in the light propagation direction y, an interaction between the light and the acoustic wave takes place in the crystal.

ω_{ex}, ω_{or} and ω_a being, respectively, the incident optical angular velocity ('ex' for extraordinary), the emerging optical angular velocity ('or' for ordinary) and the angular velocity (the angular velocity is related to the frequency ν by $\nu = \omega / 2\pi$), we obtain:

Incident optical wave: $\hat{E}z(yt) = [Ez(y)/2] \exp j(\omega_{ex}t + k_{ex}y)$

Acoustic wave: $\hat{S}(y,t) = \left[\dfrac{S(y)}{2}\right] \exp j(\omega_a t - k_a y)$

The acoustic wave interacts with the incident optical wave to produce forcing optical waves at pulsations $\omega_{ex} + \omega_a$ and $\omega_{ex} - \omega_a$. In the lithium niobiate, $n_{or} > n_{ex}$, in case of phase matching such that $k_{ex} + k_a = k_{or}$, additive diffraction effects will be obtained such that:

$$\hat{E}x(y,t) = \left[\frac{Ex(y)}{2}\right] \exp j(\omega_{or}t + k_{or}y)$$

This corresponds to the cross polarization state and so the light passes through the analyzer if the wavelength corresponds to $k_{ex} + k_a = k_{or}$ or $\lambda = V_s(n_{or} - n_{ex}) / f_0$ where f_0 is the acoustic frequency and V_s the acoustic velocity in the substrate; λ is tuneable with f_0. At the beginning, such filters were manufactured with small crystals and, later, on integrated optic devices. The spectral width is such that:

$$\Delta\lambda = \frac{\lambda^2}{I(n_{or} - n_{ex})}$$

Integrated filters with interaction lengths L = 4 mm with $\Delta\lambda$ = 2.5 nm were obtained [195], as well as with $\Delta\lambda$ = 1 nm [196] and [204]. Applications of acoustooptic tuneable filters are proposed for wavelength selective circuit switching and packet switching [204]. (A few other references are given in our bibliography from [195] to [204] and from [453] to [454]).

11.4 Wavelength multiplexers using polarizing beam splitters

Wavelength filtering can be carried out using polarization [205]-[210]. Several variants of such devices exist. A particularly interesting principle has been proposed by J. Charlier et al (Figure 95) [206] for optical intersatellite communication with four wavelengths in which the adjacent channels are separated using polarizing splitter cubes and retarding plates. In this way, λ_1 oriented in the plane of the drawing passes through the cube A and after a double pass through the retarding plate R_1 is rotated at $\pi/2$ from its incident polarization and is then reflected upwards by A towards cube B. This beam is multiplexed with λ_2 which is oriented in a perpendicular direction by reflection on cube B.

Another device from Masufumi Koga et al is shown in Figure 96.

Wavelength multiplexers using polarizing beam splitters

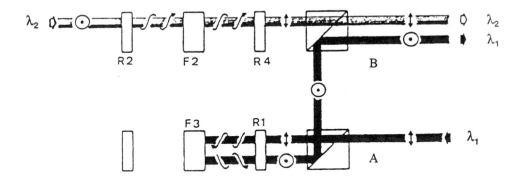

Figure 95
J. Charlier et al's multiplexer, Spie, vol. 1131, 1989 [206].

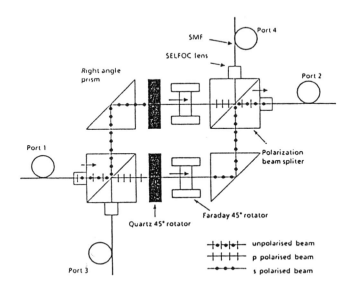

Figure 96
*Masufumi Koga et al's multiplexer
(Transactions of the IEICE, vol. E72, n°10, Oct. 1989) [209].*

11.5 Neural network multiplexers

A neural network is a structure of single elements in interaction. The concept stems from the work of biologists on brain neuron networks. Mathematical studies, statistical physics and modern information theory symbiosis have renewed interest in this concept [211]. Such artificial systems have the principal following characteristics in common with the human brain:
- learning ability,
- associative memory,
- self organization.

In practice, neural networks are currently based on connected processors or/and connected optical devices. In the future, molecular electronics may play a prominent part [212].

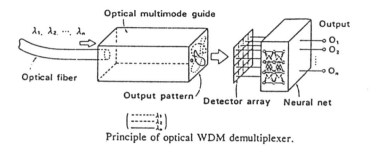

Principle of optical WDM demultiplexer.

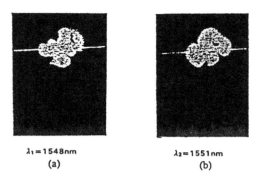

$\lambda_1 = 1548$nm
(a)

$\lambda_2 = 1551$nm
(b)

Figure 97
*Demultiplexer using a simple multimode guide
and an electrical neural network.
M. Koga et al, IEEE Photonics technology letters, vol. 2, n°7, July 1990 [213].*

M. Koga and T. Matsumoto show (Figure 97) [213] how these principles can be used for the demultiplexing of two signals at wavelengths 1548 and 1551 nm and 100 Mb/s. The device uses a multimode guide giving a spectrum-related output pattern. This pattern is analysed on a detector array followed by an electrical neural network that is reconfigurable at the transmission speed.

11.6 Photodiode array demultiplexers

Instead of discrete detectors in the grating multiplexer, a photodiode array can be used [214]-[216] and [277]. An example is given in Figure 98. Problems of electrical crosstalk between nearest receiver circuits have to be taken into account. This may limit the useful rate. In 1990 [215], the sensitivity of optical detector elements was −31 dBm at 200 Mb/s and −27 dBm at 400 Mb/s in G. J. Cannel *et al.*'s device. This sensitivity was improved in 1992 [216] to −32.1 dBm at 622 Mb/s on an eight-channel device.

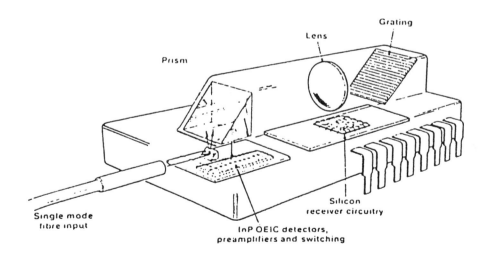

Figure 98
G. Cannel et al.'s grating demultiplexer with integrated receiver, STC.PLC, 1990 [215].

11.7 Short wavelength transmission over single-mode fibres optimized for long wavelengths

In some cases, for cost saving, it is possible to use near infrared sources such as 780 nm compact disc laser sources in fibres that are single-mode at 1.3 or 1.5 µm. Several teams have demonstrated that the problems of noise due to the resulting multimodal propagation can be tolerated in narrowband applications with self-sustained laser oscillation (for example, 66 MHz·km for trimodal propagation [217]).

The unperform'd, more gigantic than ever, advance, advance upon me
Walt Whitman

CHAPTER 12

Wavelength division multiplexing and optical amplification

12.1 Introduction

Much work has been done on semiconductor optical amplifiers, for instance in [185]. Fabry-Perot cavity amplifiers are not adapted to the simultaneous amplification of several wavelengths. But they have been used for tuneable wavelength selection. Semiconductor travelling wave amplifiers (TWAs) allow for amplification over a large wavelength range. Purely optical amplifiers have been manufactured, using Raman or Brillouin effects in single-mode fibres. However, at the present time, the solution most frequently used in commercial applications is amplification in rare earth doped single-mode fibre. In 1992, such amplifiers allowed error-free transmission at 5 Gb/s over more than 9000 km. Wavelength division multiplexing will increase these, already amazing, possibilities.

12.2 Tuneable Fabry-Perot amplifiers

In order to filter the different closely spaced wavelengths from a set of distributed feedback laser (DFL) lines successively, a Fabry-Perot cavity consisting of a semiconductor of length L, with input and output reflectivities at each extremity, R_{F1} and R_{F2} respectively, and round trip gain G_a can be used [219]-[221]. The power amplification [221] from facet to facet is:

$$A = \frac{(1 - R_{F1})(1 - R_{F2}) \exp G_a}{1 + R_{F2} \exp(2G_a) - 2R \exp(2G_a) - 2R \exp(G_a) \cos(2\beta L)}$$

with $R = \sqrt{R_{F1}R_{F2}}$ and $\beta = 2\pi\nu n / c$ (ν is the optical frequency, n the effective refractive index and c the speed of light). It can be seen that we get a high optical frequency selectivity. The selection can be obtained through temperature or injected current variation. As an example, using a semiconductor cavity for 1.5 µm with a coated facet at 90% reflectivity, and the other facet without coating, Kazovsky and Werner [221] demonstrated that a 25 dB gain can be obtained, with approximately 1.5 GHz spectral width and good adjacent channel suppression (20 dB at 5 GHz frequency offset). The device is tuneable over 150 GHz with a change in current of ±10% and a change in temperature of ±7°K. The authors predicted the possibility of expanding the tuning range to over 820 GHz with more complex structures (three-section DBR). The device is sensitive to polarization. If the state of polarization of the received signal varies with time, as is generally the case in telecommunications, this problem must be solved.

12.3 Travelling wave semi-conductor amplifiers

Braun et al [222] demonstrated that it is possible to use such an amplifier for the simultaneous amplification of ten wavelengths around 830 nm. Coquin et al [224] showed the simultaneous amplification of twenty wavelengths near 1.54 µm using a TWA with a GaInAsP heterostructure. The active length is 508 µm and an antireflection coating is used on both facets. A 6-9 dB gain is obtained if the incident signal level is larger than −32 dBm per channel. Koga and Matsumoto [225] demonstrated how the crosstalk depends on the incident power. Typically, they gave a dynamic range of 14.5 dB for twenty channels at 200 Mbit/s with an input power limited to −12 dBm.

12.4 Brillouin scattering amplifiers

Stimulated Brillouin scattering is a non-linear effect encountered in silica fibre when the power is increased (above a few milliwatts). It generates a frequency shifted light (mainly Stokes lines at lower frequencies) which provides optical gain that can amplify weak signals. The gain bandwidth is much less than the information bandwidth generally used in telecommunications. However, this effect is very useful for amplification with a narrow spectral selection and has been proposed for amplification and spectral selection. It has been demonstrated ([226]-[227]) that Brillouin scattering amplification allows the selection of channels with separations as low as 1.5 GHz and a 128-channel, 150 Mbit/s configuration was tested.

A subcarrier multiplexing optical WDM system in which the first-stage selection in WDM is based on Brillouin amplification is analysed in [460].

12.5 Raman scattering amplifiers

Raman scattering, like Brillouin scattering, converts a small portion of an incident frequency into other frequencies. The effect can be stimulated and used in order to transfer energy from a pump laser to a weak signal. In Raman scattering, the Stokes shift is much larger and the gain bandwidth is larger than in Brillouin scattering.

Thus, the simultaneous amplification of several multiplexed wavelengths is possible and was proposed in 1987 [228]. The problem of crosstalk between channels is particularly difficult to solve [229]. But crosstalk can be reduced by reducing of the channel interspacing and using sufficient power and an appropriate pumping wavelength in a 'post Raman transmitter fibre Raman amplifier'. The choice of pumping favours the shortest wavelengths in which the power depletion due to stimulated Raman scattering is expected to limit the system transmission distance [230] and [231].

12.6 Rare earth doped fibre optic amplifiers

In these devices, the signal to be amplified and an optical pump are superposed. The pump excites the doping ions: rare earth such as erbium, praseodymium and neodynium, to a higher energy level from which amplification takes place by stimulated emission. In term of energy levels, we have a 'three-level system'.

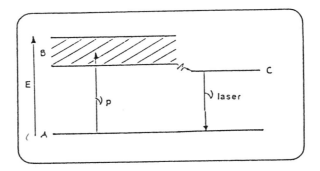

Figure 99
Energy levels E_n

ν_p: *pumping,* ν **laser** $= \dfrac{E_C - E_A}{h}$

In such amplifiers, using the pumping photons energy $h\nu_p$ (where ν_p is the optical frequency), the medium reaches the energy state B. By non-radiative energy transfer, the medium is brought to level C with a relatively long relaxation time. The transition from C to ground state corresponds to spontaneous fluorescence. If C is sufficiently well populated, a population inversion between C and A is obtained, allowing the stimulated emission phenomenon from C to A, and hence amplification can take place. Because the ground state is highly populated, the population inversion is generally more difficult to obtain than in four-level systems in which the lower level of the laser transition is not the ground state. Maiman gave the equations at equilibrium [232]. The host material is generally silica for erbium ions. However, the use of fluoride glasses as host materials has several advantages, among them a lower phonon energy than that of silica. Up to now, ZBLAN fibres are used for praseodynium (the best candidate from the standpoints of maximum gain and amplifying wavelength range at 1.3 µm) [378c].

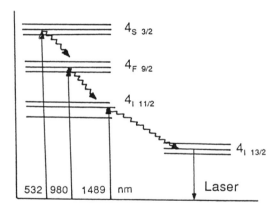

Figure 100

Transitions of erbium doped fibre laser.

Practically, pumping wavelengths at about 535, 980 or 1489 nm can be used for erbium. D. Payne *et al* [233] and [234], using a CW ('Continuous Wave') argon laser pumped at 514.5 nm and single-mode silica fibres doped with Er^{3+} obtained lasing with a threshold of 4 mW. These authors obtained the tuning range shown in Figure 101. This range can be extended by codoping: erbium, germanium, phosphorous, aluminum [235a] and/or using fluoride instead of silica fibres [235b]. J. Boggis *et al* obtained a maximum gain variation of 2.1 dB over 30 nm width at about 1541 nm. The sensitivity improvement of signal detection is 13.4 dB, giving −32.4 dBm at 5 Gbit/s [236].

W. B. Sessa *et al* [237] obtained a gain of 22 dB between the entrance and the exit of alumino silicate, erbium doped, single-mode fibre (500 ppm Er^{3+}), on a

1.5 m length with a pumping power of 200 mW at 528 nm. They were able to amplify sixteen coherent channels at 155 Mbit/s. The wavelengths are closely spaced: 10 GHz between 1539.1 and 1540.5 nm.

At the OFC'90 conference in San Francisco, an experimental demonstration of wideband optical fibre amplification was reported by Way *et al* using sixteen DFB lasers with 2 nm spacing for six channels at 622 Mbit/s and ten FM TV channels.

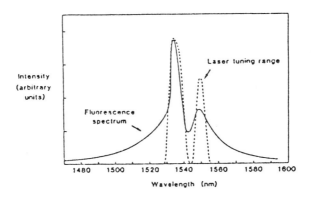

Figure 101
A typical erbium laser wavelength tuning range (silica fibre).
(Payne et al [234])

A. M. Hill *et al* [238] built a ten-wavelength, 2 Gbit/s per channel, network in the 1531-1562 nm wavelengths range. They showed, with only one of their doped fibre amplifiers, that it is possible to revolutionize telecoms networks: for example, sharing a single amplifier between 7203 customers, to distribute signals at 1.2 Gbit/s over 30 km. Such solutions led to proposals for broadband networks that are able to broadcast hundreds of broadband signals to thousands of subscribers [239] and [240]. In 1991, a field trial was conducted in Roaring Creek, Pa, by ATT Network Service Division. Four wavelengths at 17 Gbit/s were transmitted through the same 520 miles with amplifiers placed 44 miles apart [241]. Moreover, note that the fibre amplifier allows the use of the LED slicing technique (see Section 11.1) for long-distance and relatively high bit rate transmission (for example, three wavelengths at 1536, 1548, 1560 nm, 140 Mbit/s and 110 km) [242]. The advantages and drawbacks of amplifiers of this kind are as follows:

Advantages

. Excellent coupling: the amplifier medium is single-mode fibre.
. Insensitivity to light polarization state.
. Low sensitivity to temperature.

- High gain (for example in [243] 10-11 dB/mW at 0.98 µm pumping)[1].
- No distortion at high bit rates.
- Simultaneous amplification of wavelength division multiplexed signals over a 30 nm wavelength range at least.
- Immunity to crosstalk among wavelength multiplexed channels (to a large extent) [244].

Drawbacks

- Pumping laser necessary.
- Difficult to integrate with other components.
- Limited with Er^{3+} to about 1540 nm ±30 nm, but with other dopings, for example: Nd^{3+} or Pr^{3+} [378] 1300 nm becomes available.
- Need to use a tuneable gain equalizer for multistage amplification[2].

We can see that the advantages overcome the drawbacks, so frequent and systematic use is likely in multiplexed optical networks.

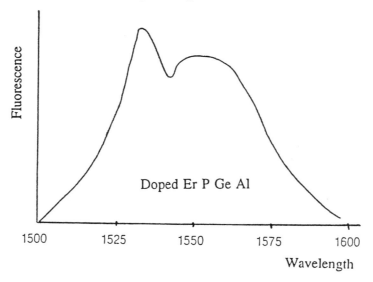

Figure 102
ErPGeAl doping fluorescence spectrum.

---

[1] At ECOC'92, R.Laming *et al* and, independently, Lumholt *et al* reported quantum-noise-limited erbium doped fibre amplifier with up to 54 dB gain is obtained with a special optical isolator between two amplifier stages.

[2] For example, a total gain imbalance of 19.3 dB has been successfully reduced to 2.3 dB by employing Mach-Zehnder-filter tuneable gain equalizers by H.Toba *et al* (ECOC'92) on 100-channel FSK signals at 622 Mbit/s in a six-cascade in-line amplifier system.

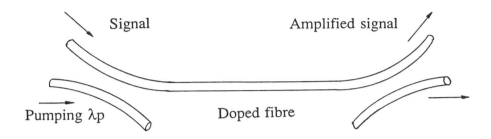

Figure 103
Principle of pumping and filtering of fibre lasers (one solution).

Remarks

The gain spectrum of an optical fibre amplifier is not flat. Hence, for long-haul systems containing several cascaded amplifiers, a small gain difference within each amplifier would result in large differences among the WDM channels. Gain equalizers are required. For example, acoustooptic tuneable filters are used to flatten the gain spectrum of erbium doped fibre amplifiers in [457]. (A theoretical analysis is given in [458]). Sm^{3+}-doped fibres may be used for gain flattening of erbium doped single-mode fibres in the range 1.53-1.57 µm [459].

Et maintenant, comme un germe de dimensions planétaires, la nappe pensante qui, sur toute étendue développe et entre-croise ses fibres, non pour les confondre et les neutraliser, mais pour les renforcer en l'unité vivante d'un seul tissu ... [1]

Pierre Teilhard de Chardin

CHAPTER 13

Application of wavelength division multiplexing to telecommunication networks

13.1 Introduction

At the present time, telecommunication networks are relatively heterogeneous. This is the consequence of an evolution, during the course of which the transmission hardware as well as the operation methods were progressively modified to meet ever-changing service requirements.

The different network types may be sorted according to their functions, their location, their routeing mode, their transfer mode, or according to their topology.

---

[1] And now, as a planetary spread germ, the thinking sheet which, on all ranges develops and interlaces its fibers not to blur and neutralize them, but to reinforce them in the living cohesion on one texture ...

Introduction

So, according to their **function**:

Networks	Typical bit rates
– Telephone network	4 kHz or 64 kbit/s
– Data network	
. Personal computer	1 Mbit/s
. Work station	100 Mbit/s
. High definition graphic mode	1 Gbit/s
– etc.	

According to their **location**:

Networks	Typical distances
– Local area network (LAN) from 10 Mbit/s (Ethernet) to 100 Mbit/s (Fiber Distributed Data Interface, FDDI)	
– Metropolitan area network (MAN)	A few km
– Interoffice network	From 10 km to about 100 km

According to their **routeing mode**:

– Broadcasting network
– Switching network (by packet or not)
– etc.

According to their **transfer mode**:

– Digital network
– Analogic network
– Frequency modulation (electrical or optical)
– etc.

Many examples could be added to each of the above categories. The transmission and, above all, the switching of a larger and larger number of signals that are increasingly diversified, and with bandwidths now larger than 1 Gbit/s for several applications, leads to a bottleneck on the lines and electronic circuits which only optical transmission and processing can relieve. Wavelength division multiplexing (WDM) is also frequently referred to as optical frequency division multiplexing (OFDM) when the wavelengths are closely spaced (spacing between channels smaller than 1 nm). It allows a linear increased capacity in number of channels, this number having its own limits (see Chapter 15: the limits of WDM and OFDM). It contributes to more flexibility in system design and, moreover, it allows new switching and routeing facilities in the optical domain. Besides, the optical multiplexer permits the evolution of existing systems and can contribute to network cost saving.

In other respects, P. S. Henry [245] showed that it is necessary to supply a minimum power level for a given bit rate to the receiver and that it is impossible to build lossless coupler power adders, which would be necessary to couple the sum of the powers emitted by the different customers' sources to the network, without using wavelength separator devices. To do so without such devices would violate the second law of thermodynamics. Of course, this is not the case with wavelength division multiplexers.

Figure 104a
Electronic multiplexer 16:1 at 2.0 Gbit/s.
(SGS-Thomson.)

It is important to consider the ordinary time division multiplexing limits. This technique will be used primarily for medium bit rates and in association with

Introduction

wavelength division multiplexing for high bit rates, and in some other cases. In 1988, for instance, a commercialy available video multiplexer such as DG 538 or DG 536 of Siliconix (eight or sixteen channels) had a bandwidth of 300 MHz [246]. Electronic multiplexers at 4 Gbit/s built in GaAs (FET) [246] and 6 Gbit/s electronic multiplexers in standard 2 μm bipolar technology [248] were developed in several research laboratories. In this last multiplexer, the operating speed was essentially limited by the D master-slave flip-flops. But the authors pointed out that a better bipolar technology with super self-alignment would have led to 10 Gbit/s. In reference [249], 8-bit heterojunction bipolar multiplexer and demultiplexer chips operating above 6 Gbit/s were also reported. In 1989, GaAs 2.5 Gbit/s, 8-bit multiplexer/demultiplexer circuits were launched on the market [250]. To our knowledge, this was probably the ultimate limit for commercialy available components at that time.

In a passive optical network, broadband distribution could possibly use subcarrier multiplexing (SCM), that is, an electronic frequency modulation technique. The resulting signal modulates a laser. In this way, sixteen SCM channels could be used in an optical passive video distribution with a SNR of 47 dB, an optical power budget of 34 dB and a dynamic range of 19 dB [251].

P.Hill demonstrated a system using two microwave subcarriers to transmit 8 Gb/s with a RF bandwidth efficiency of 1 bit/s/Hz in coherent detection [252]. In 1992, P. A. Davies and Z. Urey showed how SCM can be achieved using analog intensity modulation of optical sources with available microwave components [253]. Ultimately, in time multiplexing, one of the problems is clock recovery. Barnsley *et al* [254] demonstrated clock recovery at 5 Gbit/s for use in a 20 Gbit/s optical time division multiplexing, and Farrell *et al* [255] demonstrated a new time synchronization method.

To the question 'Do we have to multiplex the signal in the electronic domain, in time (TDM), or in frequency (FDM), or do we have to multiplex in the wavelength domain (WDM or OFDM)?', the answer is not easy and the optimum solution is generally found in the association of the different techniques. For low bit rate services (2 Mbit/s), it is sometimes better to use TDM techniques only. For uncompressed, high-definition television broadcasting, wavelength division multiplexing is highly recommended [256]. Hereafter in this book, we will often see examples of combinations of the different techniques. It is likely that applications such as video networks linking workstations, television studio centre signal routing systems, video conference networks, interactive video training systems, bank information service networks and data transfer networks between computers, integrated service digital networks (ISDN), teledistribution, and generally all broadband networks, will use time and wavelength multiplexed optical lines increasingly [257], [258] and [259].

13.2 Physical and virtual topologies

Some usual network configurations are as follows:

Star configuration: the usual architecture for the connection between a central computer and its terminals. In telephone networks, this architecture is used for routeing bidirectional communications, service, signalling, control and supervision. In videocommunication networks, this topology is generally used between the central office, remote terminals and customers.

Tree configuration: this configuration is used between a central computer, data concentration and data terminal equipment or, in many cases, in distribution or broadcasting.

Ring configuration: this configuration can be used between several data terminals connected to a computer. The FDDI uses this topology.

Mesh configuration: in this configuration, there exists more than one path for data transmission between two points on the network, with the considerable advantages of reliability and flexibility. The main example is the telephone network.

The examples given above are not exclusive.

It is understood that a practical network is very often made up of an association of architectures that constitute the physical medium of the network between stations. In the local network, simple or double star structures are often used. According to the star coupler type used in the central node, the network is so-called active or passive. The topology is called virtual when it is concerned only with logical connections between stations. One example of an optical multiplexing application is to create virtual topologies on request. So the network configuration can be modified independently of its physical topology by modification of the emitted or received optical frequencies. An interesting example of this ability is given by Bannister and Gerla [260], who used a virtual topology adjustable through wavelength division multiplexing in their 'WON' network for the optimization of the product: propagation delay × throughput (in ms × Gbit/s). WON is a generalization of the Shuffle Net concept (see Section 13.5) It is obvious that this concept can be generalized for the control of other types of optical networks.

13.3 Passive optical networks

13.3.1 A first generation of applications

The first WDM applications to customer premise networks (CPN) appeared in the early 1980s. The first networks used multimode fibres and very often two wavelengths, 0.8 and 1.3 µm, or sometimes three or four wavelengths in WDM configurations. In France, the first experiments were undertaken by the 'Centre National d'Etude des Télécommunications' (CNET) and CIT Alcatel. Let us mention the CPN network of M. Popovics *et al* [261] which provided customers with broadband digital signals: two unidirectional TV channels (video + two sound channels) and a data channel. This network used a Stimax wavelength division multiplexer with three laser diodes at 780, 810 and 840 nm and 96 Mbit/s on the link to the subscriber, and an LED at 1270 nm and 2 Mbit/s for the low bit rate signals from the subscriber. With avalanche photodiodes (APD), the practical range was 3.5 km.

In Japan, the interactive TV networks of K. Nosu *et al* are of interest [262]. In the first network, they had: four LED wavelengths and APD receivers for three bidirectional TV channels with 4 MHz baseband signals at 1.06, 1.15 and 1.27 µm and a bidirectional data and telephone channel at 1.5 µm. In the second network, they had: digital signals at six wavelengths for four TV bidirectional channels at 4 MHz using lasers at 800, 825, 850 and 875 nm with Si PIN receivers and two data and telephone bidirectional channels using an LED at 1060 and 1300 nm with avalanche detectors.

In Germany, in the early 1980s, several customer networks using multiplexing were planned. Let us mention particularly one of the solutions tested within the German Post Office project Bigfon in Berlin by Quante GmbH [263]. This network used four channels, each at 140 Mbit/s: 830, 860, 890 nm in one direction and 800 nm in the other direction.

In France, WDM was also used early on in videocommunication networks [264]. Although the first optical fibre test-bed network in Biarritz, November 1979, still used one wavelength only (800 nm), the fibre to the home networks included provision for a second wavelength (1300 nm) to solve traffic saturation problems. In these networks, 50/125 µm GI or 85/125 µm GI fibres are used. In the Velec CGCT solution, a distributing channel is shared between subscribers using a WDM at 0.85 and 1.3 µm. This channel, from the distribution centre to two customers, provides two television channels, a tuneable HiFi and a data channel for videotex and control. Another fibre from the customer to the distribution centre is shared between eight subscribers on one wavelength only [265] and [266]. Some of the 0.85/1.3 µm duplexers could be regrouped as 'SOFT' components in order to save cost and space. In the LTT Alcatel solution, channels from and to the customers are transmitted on the same fibre and a two-channel WDM is also used [268]. However, the rate of demand for connection to the network from customers was not as high as had been anticipated. But the connection rate, however, was 100 000 per year in 1986, and, three years later, several thousand 2-channel WDMs had been installed.

In Japan, the FOBID net tested in Tokyo in April 1986 fed broadband interactive services to 3000 customers. This network uses wavelength division multiplexing at 780, 880 and 1300 nm and multidielectric filter multiplexers [269] on multimode GI fibres as the networks described previously. The system in Mikata City uses a LED source and an APD receiver at 0.78 µm (NTSC 4 MHz) and 0.88 µm (NTSC 4 MHz + 64 kbit/s), a LED source and a PIN photodiode at 1.3 µm (64 kbit/s; the maximum range is 2 km [270]). In the meantime, optical multiplexing in repeaters connecting several fibre optic passive star networks have received a great deal of attention [271]. This is another very important application of WDM.

13.3.2 Evolution towards single-mode fibre networks and towards more wavelengths

Introduction

It soon became obvious that the use of single-mode fibres was necessary to achieve the bandwidth required on high bit rate trunks. In the 1980s, the great advantages of single-mode fibres for local networks, when switching, routeing and processing of signals have to be done in the optical domain, were considered increasingly. Several groups were pursuing plans to increase the number of multiplexed wavelengths and, should the occasion arise, to use wavelength tuneable sources and receivers. A better understanding emerged of the fact that the optical networks could remain totally passive and, for instance, that switching and routeing could use the wavelength at the terminal level. Now that it has been predicted that the fibre-to-the-curb (FTTC) market will be serving 2.7 million access lines by 1995 [272], it becomes obvious that WDM is of the utmost importance.

Broadband passive optical networks with a multiwavelength star structure

Among these networks, that of BTRL, introduced in 1985, is a wavelength switched local network [273]. The network terminals are interconnected through a central star coupler. A specific wavelength emitter, as well as a tuneable wavelength filter receiver, is allocated to each terminal. A network control station operates at its own allocated wavelength, polling each terminal in turn. In the free state, the stations are tuned to the control wavelength. Polling and selecting any terminal on a given optical wavelength is under the control of this station. Other versions exist in which each terminal has tuneable rather than fixed wavelength optical source.

In the *Lambdanet* network of Bell Communication Research [274], sixteen stations are interconnected through a 'transparent', single-mode, optical fibre star network. A 16 × 16 star coupler is used; each station emits its own wavelength and receives all the other wavelengths on a Stimax or equivalent WDM (sixteen channels, with 2 nm spacing between adjacent channels). The bandwidth × distance product is 1.56 Tbit s^{-1} km.

Within the *EEC RACE* Programme, is the WTDM (Wavelength Time Division Multiplexing) managed by BBC (UK). In the first phase, the group includes NTE, BNRE, GEC-Marconi, Swindon Silicon Systems (UK), ISA Jobin Yvon, Thomson CSF-LER, SGS Thomson (F), Research Neher Lab. (NL) and Alcatel SESA (S). This project is another application of a passive star network to the development of a Broadband Customer Premises Network (BCPN), for the internal routeing network for a television studio centre. The first phase has been carried through successfully. Sixteen wavelengths, with a spacing of 4 nm and at 2.4 Gbit/s each, were specified. Sixteen-channel demultiplexers with entrance and exit fibres (see Section 8.2 and Figures 45a, 104b and 105) were used at the maximum bit rate. A diode array receiver WDM option from BNRE exists (see Section 11.6 and Figure 98), as well as integrated optic WDMs options from Thomson LCR and GEC-Marconi. These device performances are illustrated in [275] and [276]. The transmission distance of the WTDM network has been extended: 6 × 2.5 Gbit/s WDM signals were successfully transmitted over a 35 km length commercially-installed 'dark fibre', using an erbium-doped amplifier [472]. The performances of another diode array demultiplexer with four channels, studied outside this programme, can be found in [277]: at $\lambda = 1300$ nm, its sensitivity is -25 dBm at 2.2 Gbit/s, and the spacing between channels is 20 nm. A similar TV studio network has also been built in Japan [278].

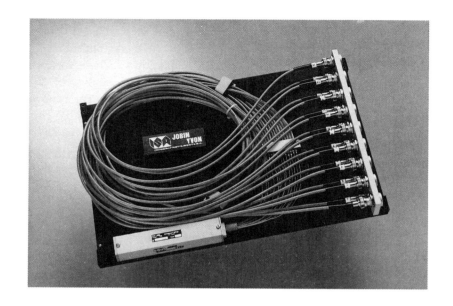

Figure 104b
*All fibre sixteen-wavelength division demultiplexer of WTDM RACE 1036-2001 network.
(With permission of ISA Jobin Yvon.)*

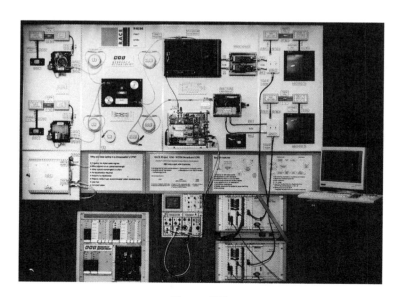

Figure 105
*Demonstration test bed of WTDM RACE 1036 network in 1991.
(With permission of the BBC.)*

Bell Communication Research's Passive Photonic Loop (PPL) network

This network uses a double passive star of single-mode fibres between the central office (CO) and the remote terminals (RT) [279]. In such a network, the switched services to the different subscribers are multiplexed on a fibre installed between the CO and the RTs. The signals are switched and routed to their intended destinations in a terminal. In the direction RT to CO, the reverse occurs. These functions can be fulfilled using time multiplexing exclusively, but this leads to the need for sophisticated electronics in the RT. Wavelength division multiplexing eliminates this drawback and makes it possible to build such a network with passive optical components, from which we obtain the acronym PPL for Passive Photonic Loop. Wagner *et al* demonstrated this using a large number of channels on the multiplexer and demultiplexer (single-mode Stimax twenty channels, 2 nm spacing between channels in the experiment), showing how it is possible to fulfil the necessary functions using ten wavelengths from CO to RT and the other ten wavelengths from RT to CO. The distance from CO to RT is 9.7 km and from RT to the subscribers it is 2 to 3 km. The bit rate is 600 Mbit/s or 1.2 Gbit/s with less than 10^{-9} BER[2] when DFB lasers are used. But the device can also work with LEDs up to 1.5 Mbit/s!

British Telecom's PON, TPON and BPON

PON, TPON and BPON are acronyms: PON for Passive Optical Network, T for Telephone, B for Broadband. These networks use a single-mode passive tree structure between the exchange and the customer terminals.

In this network, first, the infrastructure for the telephone using one wavelength only is installed. Prior to any other consideration, the economic justification takes into account the basis of known revenue earnings from telephone services, but the network is prepared for a later requirement for broadband services that will use other wavelengths. These networks fully illustrate one of the most interesting characteristics of wavelength division multiplexing: it allows network evolution without question about the infrastructure [280] [281]. A wavelength filter is included in the customer equipment (FWHM = 15 nm, $\lambda = 1.3$ µm) for the selection of the telephone service. The other services, and in particular the broadband services, can be added on the other wavelengths later on. In these networks, the time multiplexed signals at 20 Mbit/s are broadcast from the exchange to all customers (typically 128) through several passive optical splitters. The whole system involves time management and switching is controlled by the customer destination encoded on the data prior to transmission. The quality of the time management is one of the key features of the system. TPON can

---

[2] BER : Bit Error Rate

automatically control the time of travel to within 1 m accuracy over a distance of 20 km and the optical pulse amplitude to an accuracy of 0.5 dB [282] [283]. Research for improvement of the components and of the system are being pursued. Monolithic integration of a distributed feedback (DFB) laser with wavelength duplexer was reported in 1991, [284] and [285], as well as a new component and other progress [286].

The customer network ACCESS RACE 1030

A European project (prime contractor: NKT with AEG, ANT, BT, CNET, GEC, SGS Thomson, SAT, Souriau, Ericson, Telefonica, Televerket and Thomson Hybrides) [287].

It is also the aim of the European cooperation programme ACCESS (Advanced Customer Connections, and Evolutionary System Strategy) to find economically justifiable solutions for residential customer fibre-optic networks that are competitive with copper networks at the beginning, and that can be used as the starting point of an integrated broadband customer network (IBCN). Within this framework, two wavelength duplexer approaches have been studied. Both solutions consist of hybrid components: ball lens or graded index lens duplexers.

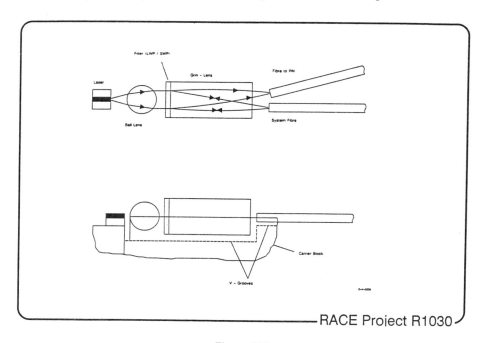

Figure 106
Multiplexer principle - Approach A.
(With permission of Access/Ellemtel.)

Passive optical networks

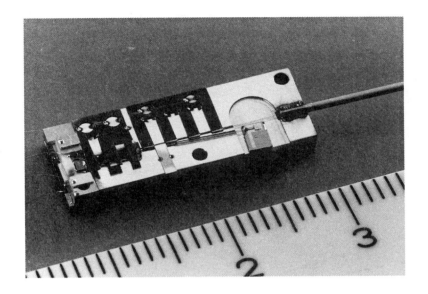

Figure 107
Multiplexer view - Approach A.
(With permission of Access/Ellemtel.)

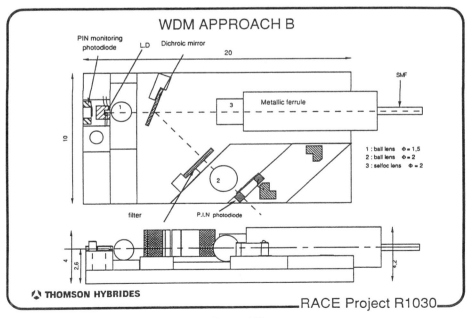

Figure 108
Multiplexer principle - Approach B.
(With permission of Access/Ellemtel.)

Figure 109
Multiplexer view - Approach B.
(With permission of Access/Ellemtel.)

Optical frequency division multiplexing (OFDM), with non-coherent detection

This technique preferably uses modulation by optical frequency shift keying (OFSK). In FDM devices, heterodyne detection is usually used, as we will see in the next chapter. However, direct detection in the optical frequency spectrum using a tuneable Fabry-Perot filter (FPF) may be used, and in this way a signal conversion from FSK to ASK (amplitude shift keying) is possible but is detrimental to sensitivity. With a double Fabry-Perot filter (for the passband of Fabry-Perot filters in series, see, for instance, [288]), the finesse and, consequently, the selectivity can be very high. In 1987, Kaminow *et al* published a tuneable Fabry-Perot filter star network. They showed theoretically that, with a double Fabry-Perot with an optical width ratio 10:11 having a finesse of 3000, up to 1000 channels can be demultiplexed [289]. It is evident that the construction of this device is delicate, but in 1989, the authors were able to verify the concept experimentally with two FSK laser channels at 600 Mbit/s [290]. Toba *et al* described a future broadband distribution network combined with switched services using a tuneable seven-stage integrated Mach-Zehnder interferometer and an FSK direct detection scheme which performed 100 channels OFDM transmission/distribution at 622 Mb/s over 50 km in 1990 [291].

13.4 Evolution towards coherent optical networks

13.4.1 Coherent detection principle

In the direct detection devices used on almost all systems described until now, the photodiodes detect the power carried by the photons and convert it into electrical current. The optical frequencies associated with the photons arriving to the receiver are distributed around a central frequency and the information that might be carried inside the frequency variation is lost. Also, the information that might be carried by the incident light phase variation is not taken into account. But the optical sources can be designed as longitudinal single-mode lasers, namely they can emit a single optical frequency with an extremely narrow spectral width ('temporal coherence'). At the receiver end, the signal emitted by such a monofrequency laser source that has been attenuated along the transmission line can be mixed with a signal emitted by another single frequency laser: the local oscillator (L_0) emitting the same optical frequency (this is homodyne detection), or with a signal at another slightly different frequency (this is heterodyne detection). The superimposition generates a signal beat, the frequency of which is equal to the source signal and local oscillator optical frequency difference. This intermediate frequency (IF) is detected in the optical receiver passband. Any amplitude modulation (AM), frequency modulation (FM) or phase modulation (PM) of the source signal is transmitted to the IF. In most cases, a binary keying on the amplitude: ASK (amplitude shift keying), on the frequency: FSK (frequency shift keying) or on the optical phase: PSK (phase shift keying) is done. For example, in FSK, the emitter may shift from one optical frequency to another close optical frequency depending on the bit 0 or 1 to be transmitted. An interesting review of an optical frequency division system is given in [463].

13.4.2 Multichannel coherent detection

The single-mode laser emits a very narrow line and so the two FSK frequencies may be very close, allowing room for many channels of the same type. This constitutes optical frequency division multiplexing (OFDM). This technique is also sometimes called high-density wavelength division multiplexing (HDWDM), the latter denomination, generally being reserved for channel interspacings about 1 nanometre, having the advantage of emphasizing that it is still wavelength division multiplexing.

13.4.3 Detectivity in coherent detection

Coherent detection corresponds to a detectivity that is better than in direct detection. If source and local oscillator are stable in amplitude and optical frequency and if the local oscillator power is sufficient, about −55 dBm in ASK heterodyne detection, −58 dBm in ASK homodyne detection and −61 dBm in PSK homodyne detection, theoretical sensitivities at 1 Gbit/s are obtained [292].

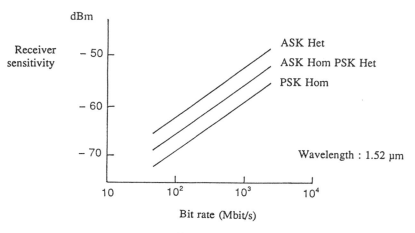

Figure 110
Theoretical sensitivity as a function of bit rate at 1.52 μm (QE = 1) using coherent techniques [292].

With coherent optical detection, the theoretical limit is given by the photon noise and in practice a sensitivity much better than that of direct detection is achieved. Moreover, the selectivity may be obtained by frequency filtering in the optical domain, for instance with the use of a tuneable laser as local oscillator.

13.4.4 Coherent subcarrier multiplexing (SCM)

Subcarrier multiplexing is in the electrical frequency domain. So it is not, strictly speaking, wavelength multiplexing. In SCM, non-coherent detection can also be used, but, in this section, we consider only coherent SCM. Thus, in the system of R. Gross *et al* [293], only one optical wavelength is used: 1.32 μm emitted by a diode pumped Nd YAg laser. This wavelength, emitted as a continuous wave, passes through a lithium niobate phase modulator, which receives the microwave signals from twenty oscillators (VCO) spread between 2 and 6 GHz with a 200 MHz spacing. A detectivity improvement over direct detection of 14 dB in FSK and of 9 dB in FM (frequency modulation) has been obtained.

13.4.5 Coherent optical multiplexing in FSK heterodyne detection

Nowadays, this is one of the most frequently used multiplexing techniques. A schematic diagram is given in Figure 111 below:

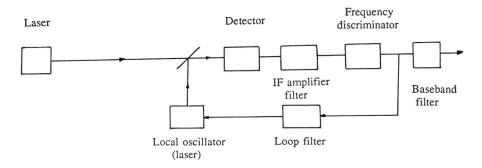

Figure 111
Schematic diagram of optical FSK heterodyne detection.

In heterodyne detection the noise results mainly from the optical noise caused by the local oscillator and from the thermal noise generated in the amplifier. The carrier to noise ratio, C/N, is improved when the power of the local oscillator is increased. The amplifier thermal noise becomes progressively negligible compared to the shot noise (Poisson process photon noise) of the local oscillator. C/N tends towards a limit. In the general case, we obtain:

$$C/N = \frac{C}{(n_s + n_c + n_t) B_{IF}}$$

and at the limit:

$$C/N = \frac{C}{n_s B_{IF}}$$

In the above formulae:
C is the intermediate frequency (IF) carrier power.
n_s is the local oscillator shot noise.
n_c is the noise arising from local oscillator amplitude fluctuation.
n_t is the amplifier thermal noise.
B_{IF} is the IF frequency filter bandwidth.

It can be shown (see, for example, [294]) that:

$$C/N_{limit} = \frac{\eta\, P_s}{h\nu\, B_{IF}}$$

In this formula:
η is the APD detector quantum efficiency.
h is Planck's constant.
ν is the optical frequency.
P_s is the incident signal power to be detected.

Optical wavelength/optical frequency conversion

Depending on one's background: the classical optical field or microwave field, one generally prefers to represent light vibrations according to their wavelengths in vacuum (the wavelength varies with the medium) or according to their frequencies (invariant of the medium). In order to evaluate the different results given in scientific papers, it is useful to be able to translate quickly.

It is well known that:

$$\lambda = \frac{c}{\nu}$$

where c is the speed of light, λ is the wavelength in vacuum and ν is the optical frequency.

Therefore,

$$\Delta\lambda = \frac{\lambda^2}{c}\, \Delta\nu$$

In practice,

$$\Delta\lambda_{nm} = \frac{\Delta\nu_{GHz}\, \lambda^2_{\mu m}}{0.3}\, 10^{-3}$$

Thus, a 10 GHz spacing at 1.5 µm wavelength is equivalent to a spacing of 0.075 nm.

There have been many experimental coherent FSK networks. In the USA, the star network of Glance et al [295], published in 1987, uses three wavelengths at about 1.28 µm spaced by 300 MHz (0.0016 nm). This network works at 45 Mbit/s with a − 61 dBm detectivity (113 photons/bit) for a 10^{-9} error rate (BER); this is 4.5 dB above the theoretical limit. It corresponds, according to the authors, to the ability to interconnect 100 000 subscribers in a 10 km range. In 1989, the same

Multiwavelength correlation routeing networks

In the star network of K. Hagishima and Y. Doi [323], the routeing addresses of data are identified by a set of wavelengths. The data signals are transmitted together on a set of wavelengths and restored by the addressee using a wavelength pattern correlation.

Shuffle Net multiwavelength bus [324] and [325]

A unidirectional optical bus connects a set of network interface units (NIU), each with two fixed emission wavelengths and two fixed receiving wavelengths. A message from any NIU to any other NIU can be addressed. In each unit, the transmitter and receiver wavelengths are necessarily different, for example, sixteen wavelengths are used for eight NIU. They are interleaved as in a shuffled pack of cards. It is possible to go from one NIU to another NIU either directly or by multihops of the data packet with its own address code. The path can be modified: in case of congestion, an alternative path of hops can be used.

Non-coherent tuneable emitter and/or receiver

Fixed receiving wavelengths can be given to the different nodes. In order to address a data message to a particular node, one can adjust a laser, the frequency of which is tuned to the wavelength of the destination node.

In such a way the FOX network [326], for instance, connects a set of N processors to a set of M memories through an optical network using double star couplers: $N \times M$ from the processors to the memories and $M \times N$ from the memories to the processors. The emitter wavelength tuning allows the addressing from whatever processor to whatever memory through the $N \times M$ coupler and from whatever memory to whatever processor through the $M \times N$ coupler. This architecture can be used for short packets of 100–200 bits that are transmitted occasionally.

In the Hypass network, [327]-[328], a double star coupler is also used for a packet switching system design. But the emitters and receivers are wavelength tuneable (with tuning response time < 5 ns and 1 ns respectively). This system is able to detect packet collisions; it has buffers and uses internal algorithms to resolve the contention problems. Note that in the Bhypass system, similar to Hypass, the use of tuneable receivers can be avoided.

Wavelength tuneable coherent receiver devices

The use of coherent detection allows us to think of making networks in which routeing can be done on a large number of channels, using dense optical frequency multiplexing [329], [317], [315], etc.]. One could replace the complex and costly current switching devices by simple, passive, non-blocking optical networks in which the light can be wavelength routed. Several experimental devices have been tested over the last few years, such as:

- **The coherent λ switch** [315], [330] and [331] which is a WTDM switch architecture using FSK detection.

- **The Bellcore Star-Track** [315], which uses an optical network with a central star coupler between N ports emitting at a fixed wavelength and receiving on coherent detectors with tuneable wavelengths.
  This network is under the control of a token generator and processor on an electronic ring linking all ports. Each input port has its own memory buffers able to temporarily store and re-emit its optical data packets under the control of the ring.

In 1989, Glance *et al* obtained a 74 photons/bit detectivity at 10^{-9} BER with six FSK channels at 200 Mbit/s [296] on a coherent star coupler network, with emitters and receivers tuneable with local oscillators. The component performances and the architectures of such coherent networks can be found in [332]-[337].

13.5.4 Remarks

1. New concepts in wavelength rearrangeable and strictly non-blocking networks have been developed by J. Sharony *et al* [338].

2. W. I. Way *et al* [339] proposed self-routeing based on subcarrier multiplexed pilot tones and acoustooptic tuneable filters in a WDM high-capacity network.

13.6 Miscellaneous point-to-point transmission links

This section is complementary to Sections 13.1 to 13.5 in which we have already presented WDM transmission lines integrated in various networks. We give hereafter other application examples of point-to-point link WDM design: bi-directional links, links associating electrical frequency multiplexing to WDM and links associating digital techniques to WDM.

13.6.1 Full duplex links

WDM is very useful for transmission in two directions on the same fibre. In such a link, an emitter and a receiver are located on each side of the line. A parasitic signal from the emitter in the vicinity of the receiver is collected by the latter. This is the near end crosstalk that must be compared to the useful signal arriving at the receiver attenuated by the line. If optical multiplexing is not used, the signal and the parasitic light have the same wavelength, so filtering is difficult. Such crosstalk may be due to defects in the Y coupler linking the transmission line to the emitter on one hand, and to the receiver on the other hand, but it is also due to all the interfaces located along the line and particularly the connectors, the reflection coefficient of which can be minimized by oblique polishing and/or by use of a matching index liquid, but it is never nil. Using a different wavelength for each transmission direction with the corresponding wavelength division multiplexers allows extensive filtering of the residual reflections. The intrinsic duplexer near-end crosstalk can be lower than −80 dB (in a Bimax component described later, for example see Figure 112). There are many examples of single fibre bidirectional links in many countries; we will only present a few cases here. In the first half of the 1980s, multimode fibres and 0.85 and 1.3 µm wavelengths were preferred. Thus, the DOTAN system, tested by the Deutsche Bundespost within the BIGFON project [340], enabled bidirectional transmissions at 284 and 81 Mbit/s on 7 km at least with an almost negligible crosstalk penalty. Afterwards, it was proposed to keep the two wavelengths at about 1300 nm and 1500 nm. In 1986, a system for the transfer of telephone and data channels from the Swiss PTT [341] operated at 1200 and 1300 nm or even around 1500 nm on single-mode fibres. In 1988, Taga *et al* built a bidirectional link at 1524 and 1548 nm, 2.4 Gb/s on each channel and on 36 km with very low crosstalk [342]. It must be pointed out that it is possible to design bidirectional links using identical wavelengths in each direction. In this way, So *et al* [343] made links at 780 nm, 2.56 Mb/s and 1300 nm, 565 Mb/s, or at 780, 890, 1300 and 1550 nm. The links were bidirectional on each wavelength. The authors gave the limits corresponding to the parasitic reflections. Thus, at 780 nm, between −10 and +50°C, with a −41 dBm signal, the permissible optical reflection limit is −51 dBm. Duplex and diplex transmissions at 34 Mbit/s over 40 km were also experimented upon in Brazil [344].

13.6.2 Links combining frequency multiplexing and/or digital techniques and wavelength division multiplexing

We have already seen many examples; hereafter, we will present complementary information on effectively installed devices.

In 1985, J. Arnaud [345] published the performances of a link with three FDM frequencies at 29, 77 and 117 MHz on 1300 and 1240 nm wavelengths. The ratio S/N is 56 dB for 13.5 km. Mendis *et al* proposed six FDM frequencies at 30, 62, 94, 126, 158 and 190 MHz on 1315 and 1230 nm with ratio S/N > 54 dB [346].

In the mid-1980s, in transmission links for the French cabled videocommunication network Artis [347], wavelengths at 0.85 and 1.3 µm were used with, on each wavelength, a TV channel (0-5 MHz) with two sound channels (6.5 and 6.9 MHz) and a stereo HiFi digital channel.

A CCETT French video link used frequency multiplexing and four LED wavelengths at 730, 810, 890 and 1300 nm on a 100 × 140 µm fibre [348]-[349]. On each wavelength, except 730 nm, there are frequency multiplexed channels centred at 19 and 46 MHz. The S/N ratio is 54 dB for a one kilometre link.

Note that complete systems using frequency multiplexing of electrical carriers associated with optical multiplexing on single-mode fibres at 1300 and 1550 nm are now on the market. These systems use multidielectric filter multiplexers such as those shown in Figure 112. They are mainly used on the main transmission trunks or videocommunication networks and on specialized point-to-point links up to 90 km [350].

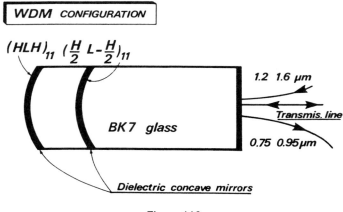

Figure 112
Bimax multiplexer.
(With permission of Jobin-Yvon.)

In the interexchange optical fibre network of Ile de France [351], FM modulation transmission has been used since 1985 and a provision was made to allow optical multiplexing at 0.85 and 1.3 µm to increase the number of television channels and to provide specialized links. Other applications of wavelength division multiplexing in transmission systems can be found in [352] (12 km, 96 Mb/s per channel, at 1200, 1300, 830 and 870 nm) and in [353] (70 km, 4 × 140 Mb/s at 1300 and 1550 nm).

13.6.3 Evolution of wavelength division multiplexing for trunk lines

Wavelength division multiplexing gains more and more applications. It has already been used on intercity trunks in the United States, in Asian countries and in Europe (1.3/1.5 µm) for many years. In 1989, British Telecom demonstrated its usefulness in submarine cable: between the Cumbrian Coast and the Isle of Man, over 94 km, operating without generators at 1525, 1536, 1546 and 1557 nm (140 Mb/s on one channel and 565 Mb/s on every other channel) [354]. In earlier chapters, we have already seen the current progress in the domain of coherent detection associated with OFDM. For long-distance links, the technique is highly interesting. Illustrating again the application of WDM, in 1991, H. Tsushima et al depicted a transmission at 32 channels, with 0.08 nm spacing centred about 1543 nm at 1.244 Gb/s on each channel, over 121 km in CPFSK (continuous phase frequency shift keying) [355]. In the last few years, there has also been substantial progress on the following components: acousto-optic demultiplexing (for instance [356]), combination of WDM with optical amplification (for instance, [357]), WDM with integration of several wavelength tuneable sources (for instance, [358]). Wavelength division multiplexing (OFDM included) is one answer to new communication needs. We have also seen multiplexed studio networks able to carry high-definition television signals (HDTV) in the RACE 1036 and RACE 2001 WTDM programmes. These techniques are also applicable to the transmission of these signals over very long distances, see [359], [360], etc. WDM is compatible with optical amplification and soliton propagation.

13.6.4 Miscellaneous remarks

- 'Wavelength dilated' switches, [464] and [465], can relax the crosstalk requirement of wavelength routeing devices.

- Wavelength conversion laser diodes can be applied to wavelength division photonic crossconnect nodes with multistage configuration [466].

In 1992, the channel capacity of a 10 km subcarrier-multiplexed lightwave system was expanded by using 1.3/1.55 µm multiplexing of 52 high-band channels (445.25–721.25 MHz at 1.3 µm) and 60 lower-band channels (55.25–439.25 MHz at 1.55 µm) [467].

The work of thinking cannot be delegated
Henry Ford

CHAPTER 14

Other applications of wavelength division multiplexing

14.1 LED linewidth narrowing for reduction of chromatic dispersion

When light emitting diodes (LEDs) are used, the transmission distance at a given bit rate often happens to be limited by the chromatic dispersion, itself linked to the source spectral width. In this way, in Paris' transmission network described in [351], the range of LEDs at 0.85 µm is limited by this dispersion to 1.5 km: therefore, five repeaters are necessary to increase the range to the 10 km necessary in some places. This problem is less important in the 1300 nm window. A standard single-mode fibre has a material dispersion zero around 1315 nm (< 4 ps/km specified at this wavelength). To simplify, with digital transmission, let us say that the dispersion effects (see Section 1.1.3) become significant when the time dispersion is 20% of the period of the signal: at 600 Mb/s, with a spectral width of 5 nm, around 1300 nm, there will be no problem up to 15 km. The problem becomes more important at 1550 nm where the dispersion is typically 18 ps/km [360]. A detailed analysis of the relations between the diode spectral width and the bandwidth in MHz of multimode fibres was published by J. J. Refi [361] in the 800 and 1300 nm wavelength range. (For example, the bandwidth is 74 MHz at λ = 1287 nm, with LEDs having a spectral width at half-maximum of 134 nm, for 1 km of 85 µm GI fibre.) The source spectral width can be reduced by the multiplexer and/or by the demultiplexer. This is particularly easy with a grating multiplexer. Indeed, in this case, the width at half-maximum of wavelength transmission function is perfectly controlled, the calculation being in good agreement with the experimental results; thus, the necessary source spectral filtering can be carried about. Stern *et al* [362] reduced the dispersion penalty of a 15 km bidirectional link on single-mode fibre using LEDs at 1300 and 1550 nm. The dispersion penalty is reduced from 1.0 dB to 0.2 dB in the 1300 nm channel at 140 Mb/s and from 3.6 dB to 2.7 dB in the 1500 nm channel at 45 Mb/s.

14.2 Transmission protection

In a wavelength multiplexed network, the provision of some protection channels using extra wavelengths can be implemented. In such a way, for instance in a passive star multiwavelength network, a tuneable laser input able to replace any other source at a different wavelength that may fail can be provided [363]. It is evident that fibre route protection using optical switches is compatible with WDM [364].

14.3 Optical interconnection

Increasingly, parallel processor architectures require the use of high-speed backplane connections. The use of optical transmissions allows bandwidth gains [365], [366] and [367]. For instance, in the USA, the developments of the Mitre Corporation and in Europe those of Thomson CSF are well advanced in this area. LED spectral slicing is particularly interesting in such applications. This technology is compatible with multichannel interconnections at more than 1 Gbit/s [368]. Among the different techniques, volume holography for chip-to-chip, board-to-board and processor-to-processor multiwavelength WDM interconnection has good potential [369].

14.4 Wavelength division multiplexers for optical pump coupling and amplified signal filtering in doped fibre lasers

We have already seen the corresponding scheme in Figure 103. For filtering, one of the channels of a classical demultiplexer with an integrated grating structure on a fibre can be used [370]. This filtering can be tuned with a rotating interference filter, a rotating grating or a tuneable Fabry-Perot [371], [372] and [373]. The multiplexers used for pumping are as follows:

- a duplexer (532-1540 nm) or (980-1540 nm) or (1489-1540 nm) for Er^{3+} doped fibre lasers;
- a duplexer (1017-1310 nm) for Pr^{3+} doped fibre lasers, [377] and [378], emitting in the 1.3 µm wavelength range.

One of the main limitations of the fibre amplifier is related to the amplified spontaneous emission (ASE). In a high-gain, copropagating amplifier, the backward propagation ASE becomes sufficiently high to deplete the pump power and reduce the population inversion mainly in the input section of the active fibre. This induces a lower gain efficiency and higher noise level. However, an optical spectral isolator may be inserted at a particular point of the active fibre which attenuates the propagating ASE much more than the pump [373]. Such isolators can be inserted between different sections of active fibres. The use of WDMs can improve these configurations, the light signal passing through the isolator being separated from the pump bypassed down to the next section of the amplifier [374] and [375]. Other limitations of the fibre amplifier are related to the gain spectrum width and flatness. The problem becomes very important in cascaded amplifiers used in WDM undersea transmission, as a small difference in gain spectrum in each amplifier becomes large after several amplifications. Thus, gain equalizers are necessary. This is particularly important in soliton propagation in which the pulse power must be controlled. These equalizers are acousto-optic tuneable filters (Shing Fong Su et al, ECOC'92), tuneable Mach-Zehnder filters (H. Toba et al, ECOC'92) or other spectrum filters. For example, A. V. Belov et al proposed a Sm^{3+} doped passive fibre filter for gain-flattening of an erbium laser (see also ECOC'92 proceedings).

14.5 Dispersion measurements

Wavelength division multiplexers are very useful for chromatic dispersion measurements of a fibre link. Among the three following measurement techniques: pulse delay technique, modulation phase technique and swept frequency technique, the second has the best dynamic range and a high accuracy (better than 0.1 ps/(nm·km). However, it requires time if the dispersion at different wavelengths is required: a reference laser and a measurement laser have to be connected again for each wavelength. The use of wavelength division multiplexing eliminates the need for the reference laser and requires only one connector, with corresponding saving in time [379] and [380].

14.6 Sensors

The fibre optic sensor signal may be wavelength encoded in different ways: back-reflection from one fibre to another fibre or in the same fibre after diffraction on a grating (the back-wavelength being associated with the grating rotation angle or with a shift in the spectral plane, [381] and [382]), back-reflection after passing through a multidielectric filter plate that is variable from one edge to the other (measurement of the position variation of the plate), wavelength variation through a Fabry-Perot whose optical thickness varies, etc. The different sensors can be frequency, time or wavelength multiplexed (see next section).

14.7 Industrial control and sensor networks

We will provide some characteristic examples. Two-wavelength multiplexing was used for measuring railway track deflection relative to the sleepers. In B. Jones' example, [383] and [384], a filter with two adjacent surfaces for different wavelengths is fixed to the railway track. If there is a displacement of the railway track corresponding to the passage of the wheels, the filter follows the displacement. The two wavelengths are alternately emitted towards the double filter through an optical fibre. The signal after passage through the double filter is collected by an exit fibre. The two wavelengths are spatially separated by a dichroic filter at the exit. The difference in the intensities transmitted at the two wavelengths is measured and the variation is related to the rail displacement.

Multiplexing using LED spectral slicing (up to 50 wavelengths per LED) for a fibre-optic process control network was developed by Framatome and Instruments SA JY in France, initially for the control of nuclear boilers in an all-optical acquisition and transmission system. In this application, the essential advantage of wavelength division multiplexing is related to the possibility of transmission through the reactor building with a single optical fibre which transmits the multiplexed information of many sensors (20 to 50 items of information on the state of valves, for instance). To our knowledge, at the time of writing, such a multiplexed network has not yet been submitted to all the developments necessary to allow its systematic use in the nuclear field but its applications to the control of industrial process are effective (oil refinery, steam generating station, etc.) [385].

Boeing, as well as NASA, Litton, Smith industries, and some administrations in the USA have studied wavelength division multiplexing for sensor networks [386] and [387]. One of the first applications was in the mobile area (ships, planes, vehicles, etc.) In France, in 1986, PSA, in cooperation with Acome-Sysoptique and Instruments SA JY, studied applications of wavelength division multiplexing to car control networks, a multiplexer in a Stimax two-dimensional structure (plastic multimode optical guide) being tested. Six wavelengths in the visible range with 30 nm spacings on a ribbon of plastic fibres were used. The weight of the device is only 5.56 grammes! The small-scale model of a car (Laude, 1986), inside which such a planar multiplexer is integrated is shown in Figures (113 a–b). The exit fibre ribbon, lit up by different wavelengths coming from the back of the car, the extremity of which is placed on the roof, is visible in the photograph.

Presentation of a small-scale Peugeot 205 equipped with a multiplexer using integrated optics

Figure 113a
Fibre array visible on the roof.

Figure 113b
The plastic integrated optic multiplexer can be seen on the floor of the car.

Applications to the industrial domain [388] are extremely numerous and we could not hope to mention them all. The most usual application is that of communication networks on industrial sites that frequently use wavelength division multiplexing (for example: the Epeire network [389] at 0.85 and 1.3 µm).

14.8 Telespectroscopy

The idea of using optical fibres for remote spectroscopic measurements probably appeared in the early days of optical fibre development. Over a long period of time, classical spectrometers were used coupled with optical fibres [390]. But the spectrometers are generally adapted to the use of entrance slits (and, possibly, of exit slits). Thus, some astigmatism is tolerated provided that one of the astigmatic focal lines can be focused along the image of the entrance slit. In 1980, a new monochromator [72] was proposed which was well adapted to the use of optical fibres since it had no astigmatism. Later, other solutions were also proposed, using, for example, new holographic gratings. In fact, the wavelength division multiplexer used in telecommunication is a multichannel spectrometer and after adaptation of the wavelength channels it can also be used in telespectroscopy. In this way, the Stimax multiplexer [73] was used for polychromatic pyrometry, [391] and [392], for pollution monitoring by fluorescence in the marine environment [393] and for biomedicine [394] in spectral identification of normal and atherosclerotic arterial tissue layers.

Of interest is the use of a combination of optical fibres and a charge coupled device (CCD) that consists of a matrix of detectors [395]. Many spectra can be recorded simultaneously (see also 'SOFT WDM', Section 11.2). Using the WDM techniques, the optics of such a device were designed for IFREMER (a French organization for marine research) with which M. Lehaître and M. Birot built an instrument for monitoring pollution (Figures 113c–116). The entrance slit is replaced by an array of 50 fibres 200/300 µm. Fifty spectra can be recorded simultaneously.

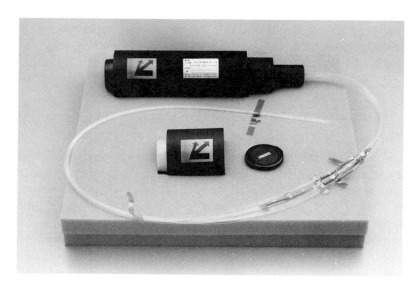

Figure 113c
*CCD spectrometer-multiplexer, optical part (Laude, 1991).
Document ISA Jobin Yvon.*

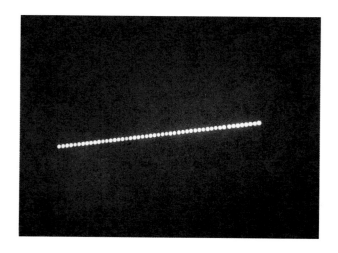

Figure 114
Cross section of the entrance optical fibre array.

Telespectroscopy 159

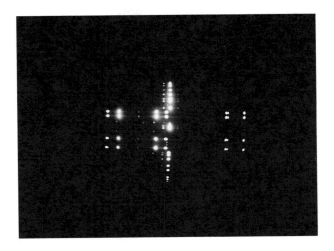

Figure 115
Spectra of mercury and sodium at different entrance fibres.

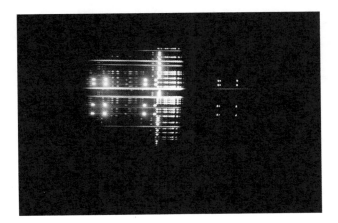

Figure 116
Spectra of tungsten, mercury, sodium and neon at different entrance fibres.

The spectral domain is about 350 to 860 nm. It is only limited by the transmission of the optics in the ultraviolet and by the size of the CCD (diameter: 18 mm) in the infrared. This range can be enlarged. The dispersion in the CCD matrix plane is $\Delta x/\Delta \lambda = 30\,000$. The resolution depends only on the diameter of the fibres used. With 200×300 µm fibres, the resolution is 2 nm in the whole field; it would be 0.1 nm with single-mode fibres. Such an instrument, if adapted to the wavelengths used in telecommunication, would allow the signals coming from 50 (even some hundreds) of different fibres to be demultiplexed with several thousands of channels per entrance. In practice, this number is limited by the CCD characteristics, the device being almost stigmatic on a field of 36×36 mm. Another limit is related to the cumulative crosstalk.

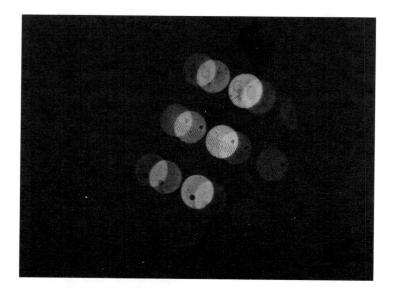

Figure 117
Somewhere in the spectral exit plane, three fibres being illuminated by a neon lamp. Three corresponding spectra of lines are seen locally.

Figure 118
Resolution test on the mercury doublet (576.96-579.06 nm).
The entrance is on any fibre among the 50 entrance fibres.

14.9 Multiplexing of radar signals

The use of optical fibres and of optical multiplexing for linking remote radar networks to their processing centre is advantageous. Optical fibre and wavelength division multiplexing is compatible with all kinds of analog and digital signals. A multichannel bidirectional link can be achieved on a single fibre with saving on the cabling and connection. Thus, on an airport surveillance radar such as the AN/GPN 12 of Texas Instruments, Leonard and Vidula [396] were able to multiplex the frequencies of three video signals (normal + MTI + Beacon) on one wavelength λ_1 and to use time division multiplexing on Azimuth Change Pulses (ACP) and Azimuth Pulses Reference (ARP) at λ_2, as well as on two voice signals at λ_3 and two control signals at λ_4. The $\lambda_1 + \lambda_2 + \lambda_3 + \lambda_4$ sum is multiplexed in wavelengths with a grating device into one fibre. Another fibre carries voice and control signals from the operation shelter (OPS) to the radar, multiplexing on two wavelengths using a dichroic filter component. The same authors give other examples of single-mode links at ten or twelve wavelengths. The benefit of using wavelength division multiplexing on phased-array radars (active antenna consisting of an active module matrix with several interconnected pairs of microwave transmitters and receivers located one beside the other) was pointed out by Deborgies and Richin [397] and Goutzoulis, Davies and Zomp [469]. These last authors also demonstrated how WDM allows the simultaneous transmission of wide-band radio-frequency and local oscillator signals, each on one wavelength, for ultra-high frequency microwave links [470].

14.10 All optical image transmission

Several years ago the idea of coding the image elementary 'points' by wavelength for their transmission through a fibre received a great deal of attention. However, to our knowledge, up to now, this has not been used on commercial devices despite the great interest shown as it allows optical transmission of an image on one fibre only. New sources [398] and WDM may renew such research.

Le temps est ce qui empêche que tout soit donné d'un seul coup.[1]

Henri Bergson

CHAPTER 15

Some limitations of wavelength division multiplexing

15.1 Crosstalk effects

In digital and direct detection systems, the penalty due to crosstalk can be determined by the signal-to-noise variations corresponding to the addition of crosstalk. This penalty can be quantified by the eye aperture in the 'eye pattern'. A given bit error corresponds to a given aperture. In the presence of crosstalk, the restoration of the eye diagram may sometimes be achieved with a signal amplification. The power amplification measured in dB, necessary for this operation, is called crosstalk penalty. The penalty essentially depends on the receiver eye and on the regeneration electronics (mainly hysteresis). In particular, in PIN receivers, thermal noise is dominant, while, in avalanche receivers (APD), the noise is quantum noise, proportional to the square root of the flux received. Quantum noise limited receivers are affected by crosstalk faster than are thermal noise receivers. In so far as the crosstalk coming from the different multiplexed channels is not correlated between them, the penalty does not increase linearly with the number of channels, but, fortunately less. Thus, some crosstalk between channels may be tolerated. Rosher and Hunwicks theoretically and experimentally show that, in LED spectral slicing, the channels can have a small overlap: −15 dB intrinsic crosstalk of each channel on the neighbouring channels does not yield a significant performance deterioration on a 34 Mbit/s device with PINFET receivers and ten spectral channels [399]. In agreement with these results, A. A. Al Orainy and J. J. O'Reilly [400] give signal to crosstalk values (SIR: Signal to Interference Ratio) in direct detection on WDM systems, for example:
SIR = 15.2 dB with PIN receiver and SIR = 16.4 dB with APD receiver with a

---

[1] Time is what excludes all to be given at once.

gain of 10, in order to retain a 10^{-9} bit error rate (BER) with a NRZ format (Non-Return to Zero). Analysing a bidirectional link at 622 Mb/s, 1300 nm and 155 Mb/s, 1550 nm, S. Geckeler found SIR = 15 dB in the worst case (622 Mb/s, RZ format) [401]. We have seen that crosstalk due to grating multiplexer defects is generally between −30 and −50 dB and does not depend too much on the wavelength spacing. For two-channel dielectric filters with a large distance between channels (850, 1300 nm, for instance), the crosstalk is −30 to −80 dB. So, in general, this crosstalk is not heavily penalized. For Fabry-Perot devices, the crosstalk is larger but it can be optimized [402] and [403].

15.2 Polarization effects in multi/demultiplexers

For single-mode links using standard fibres, that is to say without polarization maintaining fibres, the components' polarization results in selective noise. This polarization noise must be avoided or compensated for (with diversity polarization receivers or similar). In any case, the polarization penalty is directly related to the component polarization dependency. The problem is particularly important in coherent detection. The polarization dependency of grating or multidielectric filter multiplexers can easily be reduced below 0.5 dB. This dependency does not change very much with temperature, which is not, unfortunately, the case for classical fused fibre couplers. Polarization maintaining biconic couplers, with low sensitivity to temperature variations, can be manufactured [404].

15.3 Crosstalk due to Raman conversion

This crosstalk is one of the main limitations in optical multiplexing. The fibre, excited at a given wavelength, re-emits a scattering spectrum at the same wavelength (Rayleigh scattering), which is not a problem, but also at other wavelengths, longer or shorter than the excitation wavelength, respectively Stokes and anti-Stokes Raman scattering. The Stokes spectrum only has enough intensity to become a constraint with cumulative effects along the fibre. Its maximum with silica is at 430 cm^{-1} from the excitation line; therefore, such a channel distance is the worst configuration. However, Raman crosstalk may not result in practical consequences in the most common cases. P. Niay *et al* [405], working at about 1.3 μm, with two 430 cm^{-1}-spaced channels (the worst case), for − 3 dB injected powers (in such a case, stimulated Brillouin and Raman effects are negligible) and a 100% analogic modulation rate, showed that the crosstalk is about − 30 dBm for 15 km of single-mode fibre. It could reach − 25 dB for an analogous system at 1.5 μm on 25 km in the same conditions.

S. Chi and S. C. Wang [406] calculated the limitation in bit rate × maximum length of high-density, optically multiplexed links related to Raman scattering on fibres with dispersion shifted at 1.55 µm only. With 10^{-13} Joule input pulses, they found a bit rate × maximum length of 3.27 10^5 km Gbit/s. This best case could be obtained with 0.2 nm channel spacing and 240 wavelengths. See also [407] and [471].

15.4 Crosstalk due to other non-linear effects [408]-[417]

Other limits related to non-linear effects such as stimulated Brillouin and Raman scattering (SBS, SRC) or four-wave mixing (FWM) correspond to real limitations in coherent multiplexing. For instance, Waart and Braun [413], tackling the FWM problem, demonstrated that, with a multiplexed link at 100 channels, 5 GHz spaced, where it is intended to limit the crosstalk at −20 dB, the input power must be limited from 0 to −5 dBm for 5 km, and to −10 dBm for 100 km. FWM generally remains one of the main problems. But for a larger number of channels (about 300 and above), SRS becomes the dominant launch power limitation (−5 to −10 dBm). The launch power limitation due to SBS does not depend on the number of channels and is more than a few dBm in typical applications [414]. The effect of fibre chromatic dispersion on FWM is critical [417] and [418]: a negligible penalty with conventional step index fibre may become high with dispersion shifted fibres. D. A. Cleland *et al* showed that FWM is the dominant consideration for the determination of channel spacing in long-distance links (typically 560 km, 2.4 Gb/s) [419].

15.5 Minimum channel spacing related to 'uncertainty' relationships

We start by considering the optical frequency of one channel $v_0 = \omega_0/2\pi$ which is perfectly defined: it is a 'perfectly' monochromatic continuous wave source, which means that its emission lasts ceaselessly from the beginning to the end of time. An electro-optic shutter is placed on the light path, it allows light through it from $t = -\Delta\tau/2$ to $t = +\Delta\tau/2$. Let us show that after passing through the shutter, the spectrum is no longer monochromatic. As a matter of fact, a square function, of $\sqrt{\pi/2}$ height, for $|t| < \Delta\tau/2$ in the time domain, corresponds to a Fourier transform $F(\omega) = \sin((\Delta\tau/2)\omega)/\omega$ in the frequency domain. This is obtained, knowing that $F(\omega) = \frac{1}{\sqrt{2\pi}} \int_{-\infty}^{+\infty} f(t)(\exp-i\omega t)dt$, considering the usual validity conditions that are satisfied here. The $\Delta\omega$ frequency width, from ω_0 to the first zero minimum of $F(\omega)$ is such as $\Delta\omega = 2\pi/\Delta\tau$; this leads to the relation: $\Delta\nu\,\Delta\tau = 1$. In this way, a frequency distance limit, under which it is impossible to set the frequency multiplexed channels closer, corresponds to a given information rate per channel. For example, for $\Delta\tau = 0.4\ 10^{-9}$s elementary duration light bits, the limit is $\Delta\nu \approx 2.5$ GHz, corresponding to a minimum distance between channels of $\Delta\lambda \approx 0.02$ nm at a wavelength of 1.5 µm. The same relations applied to Gaussian pulses with Fourier transform calculated as in Section 11.1.2 gives $\Delta\nu\,\Delta\tau = 1/\pi$ where $\Delta\nu$ and $\Delta\tau$ are, respectively, the optical frequency bandwidth and the pulse width at 1/e.

In practice, the signals are not single and ideal pulses and, when crosstalk specifications are taken into account, it is generally necessary to use channel spacings several times larger than given by these 'uncertainty' laws. For example, with 150 Mbit/s in FSK-NRZ[2], the optical channel spacing which results in a 1 dB crosstalk penalty is 1.8 GHz when the source linewidth is negligible and 4 GHz when IF[3] and laser linewidths are 50 and 25 MHz respectively [439]. This is twelve times the limit given by the above formula in the first case. Another example, the channel spacing used in [295], Section 13.4.5, is four times this limit. In the multiplexing of solitons experiment by Mollenauer *et al* [423], Section 15.8, with Gaussian pulses of 40 ps FWHM, the channel spacing of 0.36 nm is 6.8 times the limit given by $\Delta\tau\,\Delta\nu = 1/\pi$.

---

[2] FSK-NRZ: frequency shift keying - non return to zero.

[3] IF: intermediate frequency.

15.6 Effect of spectral filtering on multimode lasers

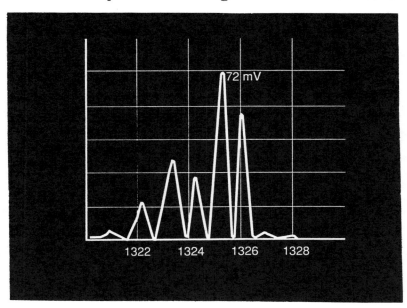

Figure 119
Typical multimode laser spectrum.

The emission spectrum of a multimode laser consists of a set of wavelength peaks spread over a few nanometres, as shown in Figure 119. Integrated over all wavelengths, the emittted intensity is unchanging if the current intensity is constant. If the laser is fed with square waveform current, it emits square waveform flux. The signal from the detector receiving this flux is of high quality (Figure 120).

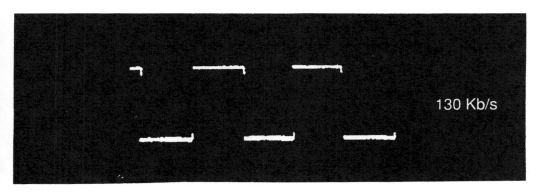

Figure 120
Output signal of a detector collecting all the flux emitted by the multimode laser.

When an optical demultiplexer is inserted between the laser and the receiver, high-frequency noise may be generated (at a few GHz) if the corresponding demultiplexer spectral transmission width is similar to the laser spectrum width or smaller. For instance, when the laser of Figure 119 was filtered through a demultiplexer, on a channel centred at 1325 nm, with 2.3 nm spectral width (FWHM) and 1 nm at 90%, the signal received was highly perturbed by high-frequency noise, whatever the modulation frequency (see Figure 121).

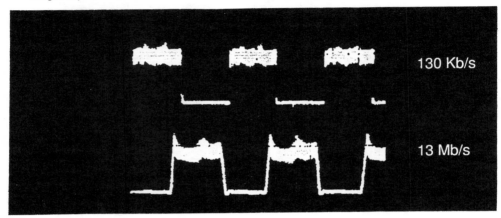

Figure 121
Detector output signal after demultiplexing in 1987.
(Recording at ATT Bell MH at the output of a Stimax demultiplexer (2.3 nm FWHM).)

This noise is the result of mode hopping at high frequency, which is revealed by the spectral filtering. Consequently, it is necessary to make sure that the multiplexer channel widths are larger than the emitter spectral widths or the mode-partition must be controlled.

15.7 Directly modulated DFB LD spectral spread

In high-speed narrow wavelength spacing WDM transmission systems, using directly modulated distributed feedback laser diodes (DFB LDs), it is necessary to take into account the spectral spread induced by the modulation of the laser. M. Henmi *et al* [438], checking time-resolved spectra of such sources, observed two peak spectra. Under a few Gbit/s modulation, the wavelength difference between the first and the other pulses was as large as 0.5 nm and the spectral spread for the first pulse was as large as 0.7 nm. Under certain assumptions, with Gaussian optical pulses in large-signal modulation conditions, it was shown that the dynamic spectral broadening Δv of 'dynamic single-mode' lasers directly modulated at high frequencies, can be approximated by $\Delta v \Delta \tau = \dfrac{2 \ln 2}{\pi} \alpha$ [473]-[475], where Δv is the dynamic frequency spectrum FWHM of a single longitudinal mode, $\Delta \tau$ is the FWHM of a Gaussian pulse and α is a coefficient (typically, 2 to 8). It is worth comparing this formula with the limit formula in Section 15.5. For non-Gaussian pulses, the chirp is larger. So, external modulations or complicated laser structures must be employed for very high-density optical multiplexing. However, laser structures able to retain a small spectral width (typically 100 kHz) under frequency modulation have become available. In 1992, Dr Okai proposed a CPM-MQW-DFB[4] laser with a frequency width of 56 kHz over 13 nm wavelength modulations, and Dr Sherlock *et al* demonstrated a DFB laser with a flat FM response in excess of 2 GHz/mA with an associated frequency bandwidth in excess of 5 GHz [444] This device was continuously tuneable over almost 2 nm without mode hopping.

15.8 Wavelength division multiplexing of solitons

We have seen some limitations resulting from material non-linearity. But non-linearity is not necessarily a drawback: in Chapter 1, we pointed out the attractive feature of soliton propagation, which in fact uses non-linearity for long-distance transmissions. Ultra long-distance WDM solitons using broadband fibre amplifiers are now being studied in several laboratories, such as [420]. It was proposed to use the flux of non-dispersive solitons, [421] and [422], with loss compensated by the Raman gain obtained by periodic injection of cw pump power to obtain ultra-long-distance high bit rate for all optical transmissions (50 ps solitons over more than 6 000 km in 1989 [27]). In 1992, error-free soliton transmission over more than 11 000 km at 10 Gbit/s in WDM at 1555.32 and 1555.68 nm was demonstrated [423]. It was shown, [422] and [424], that with uniformly cancelled loss, solitons with different wavelengths emerge from mutual collisions without modification. Mollenauer *et al* [425] showed that in lossless and unperturbed conditions, the only result after collision is a time displacement between solitons of different frequencies v and $v + \delta v$ such that of $\delta t = \pm\ 0.1786\ 1/\tau\ (\Delta v)^2$, where τ is the FWHM pulse intensity. So, if we wish to keep the time shift to a given minimum, this sets a minimum allowable Δv. On the other hand, in the case of variations in loss and dispersion, a frequency shift results from the soliton collision. A perturbation length L_p being defined as the amplification period itself or being applied to other perturbations such as dispersion, and a collision length L_c corresponding to the overlap segment limited at half-power, Mollenauer *et al*

---

[4] Corrugation Pitch Modulated - Multi Quantum Well - Distributed Feed Back.

showed that if $L_c = 2 L_p$, this sets a maximum allowable $\Delta\nu$ such as:

$$\Delta\nu_{max} = 0.31 \frac{Z_{so}}{\tau L_p}$$

Z_{so} being the collision length (see Section 1.6).

In the practical case of 9 000 km, L_p = 40 km, dispersion = 1 ps/(nm·km), τ = 50 ps at 4 Gbit/s at each frequency, Z_{so} = 930 km, $\lambda \approx$ 1.55 μm, $\Delta\nu$ = 146 GHz [425]. If $\Delta t_{max} = \pm$ 7.5 ps is set, this corresponds to a spacing between channels limited to 0.3 $\langle$ $\Delta\lambda$ $\langle$ 1.2 nm. However, a frequency dependent gain obtained, for instance, with dispersion shifted distributed Er^{+3} doped fibres may be used to compensate for the soliton frequency shift [426]. Other solutions [427] may be discovered to reduce the limits down to the uncertainty limits.

From another point of view, limitations in the transmission capacity of soliton-based communications using both dispersion shifted fibres (DSF) and normal fibres are examined in [437]. This study is not specific to WDM systems. In the best case, corresponding to DSF fibres, loss limitation and collision limitation on distance are, respectively, 50 and 24 km at 50 Gbit/s. It was shown that 'the main advantage of soliton transmission is not that it removes the dispersion effects, but rather a technique to combat the ills of self and cross-phase modulation that prevents multichannel operation in linear systems' (N. A. Olsson and P. A. Andrekson, 1992).

15.9 Some other remarks

- Limitations due to optical beat interference in wavelength division, subcarrier frequency division and multiple access networks were analysed by Shankaranarayanan, Elby and Lan [428].

- It is worth being aware that in a few cases, in particular when sources are imaged on intermediate diffusers, spectral changes can be induced by the state of coherence of sources, [429] and [430]. This may explain residual crosstalk problems in specific situations.

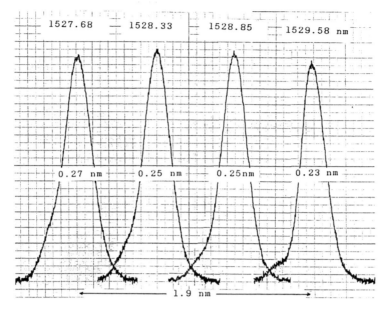

Figure 122

Transmission function of a four-channel all single-mode fibre demultiplexer with 0.6 nm channel spacing (our practical limit on Stimax device in 1992).

If you should suddenly feel the need for a lesson in humility, try forecasting the future of physics.

Daniel Kleppner

Conclusion

As the story goes, at the end of the last century, Lord Rayleigh said that physics was practically complete and that all basic ideas were known except for some constants to be measured and various small tasks. What has happened since then is known, and what crosses our mind at the moment we are about to conclude is, instead, what Thomas Jefferson said one century before that time: 'It is impossible for a man who takes a survey of what is already known, not to see what an immensity ... remains to be discovered'.

Since the first papers by Tomlinson in 1977, a great distance has been covered. We have tried to discuss some milestones, without being able to discuss them all, or to cite the names of all those who have contributed towards making wavelength division multiplexing one of the main technologies in optical telecommunication.

The 100 channels 10 GHz spaced and simultaneously amplified in an aluminum erbium doped fibre, using distributed feedback lasers at 622 Mbit/s, reported by K. Inoue in 1991, are henceforth a part of the long multiplexing story. Records are set but some margin still exists between the present results and the fundamental limits. There are indeed about 15 000 GHz of optical frequency bandwidth in each 1300 nm and 1550 nm window. With 5 Gbit/s bit rate, the uncertainty relationship gives about 5 GHz limit for optical frequency spacing. At this limit, optical crosstalk is probably catastrophic. With 50 GHz spacings, chosen for their single-mode fibre transmission capacity analysis by R.S. Vodhanel and R.E. Wagner [431], 350 channels at 1300 nm and 325 channels at 1550 nm are feasible and the crosstalk is penalized less. If new fibre development feasibility in a larger spectral domain could be added to that ([432] for example), the number of channels that could be planned is almost limitless.

Conclusion

But, as Bernard Shaw said (not literally, but in substance): The reasonable man conforms himself to the world, the unreasonable man tries to adapt the world to himself. Accordingly, any progress depends on the man who is unreasonable.

At present, the techno-economical problems have not been forgotten and the solutions in use in practical systems have few channels. In the late 1980s, two-channel solutions at 1300 nm and 1550 nm were accepted as the best techno-economical choice for customer networks [433]-[434]. But, from that time on, in passive systems a larger number of channels have increasingly become the key to cost saving in integrated broadband service networks (IBSN) [435] and are already the solution for high bit rate transport trunks, links between fast computers, and standard and high-definition television studio networks (TV and HDTV) and in many other domains. About 60% of the optical fibre market may be non-telecommunication over the next few years [436].

The enthusiasm for wavelength division multiplexing in the telecommunication research community is unquestionable. But the novel idea is frail and entrepreneurs able to pass from invention to innovation are those who prefer act to criticism.

Last but not least, as for any new technique, the future will also depend on the synthesis and educational tasks to be accomplished.

Bibliography

[1] R. T. Denton and T. S. Kinsel : "Optical multiplexing and demultiplexing", Proc. IEEE vol.56, p.146, 1958.

[2] O. E. De Lange : "Wideband optical communication systems", part 2, Frequency division multiplexing, Proc. IEEE, vol.58, p.1683, 1970.

[3] W. Tomlinson and G. Aumiller : "Optical multiplexer for multimode fiber transmission systems", Appl. Phys. Lett., vol.31, p.169, 1977.

[4] G. Daniels : "Case study : the 1992 Winter Olympics", Telecommunications, pp.54-57, Feb.1992.

[5] D. Gloge : "Weakly guiding fibers", Appl. Opt., vol.10, no.10, pp.2552-2558, 1971.

[6] D. Marcuse : "Theory of dielectric optical waveguides", J. Quant. Elect., Principles and applications (Y. H. Pao, ed.), Academic, New York, pp.60-78, 1974.

[7] A. Cullen : "Electromagnetic waves guided by dielectric. Summer school in electromagnetism", CNET Lannion (France), July 1973.

[8] H. G. Unger : "Effects of inhomogeneities in dielectric waveguides : graded-index media and local inhomogeneities", Summer school in electromagnetism, CNET Lannion, July 1973.

[9] M. Monerie : "Propagation in doubly clad single-mode fibers", IEEE J. Quant. Elect., vol.QE-18, no.4, pp.535-542, 1982.

[10] M. Monerie : "Analytical approximation for W-type propagation parameters", Elect. Lett., vol.18, no.9, pp.386-388, 1982.

[11] L. B. Jeunhomme : "Single-mode fiber optics principles and applications", Marcel Dekker Inc. New York, Bâle, 1983.

[12] S. H. Wemple : "Material dispersion in optical fiber", Appl. Opt., vol.18 (1), pp.31-35, 1979.

[13] P. Dupont : "Des fibres optiques monomodes à gaine d'indice inférieur à la silice, pourquoi ?", Opto 82 Conf. Proc., ESI Publications, pp.11-15, 1982.

[14] European Patent 0089655, Int. Standard Electric Corp., Single-mode optical fibre, 19.03.83.

Bibliography

[15] J. J. Bernard, C. Brehm, J. Y. Boniort, Ph. Dupont, J. M. Gabriagues, C. Le Sergent and M. Liegois : "Investigation of the properties of depressed inner cladding SM fibers", ECOC'82 Proc., pp.133-138, SEE Ed., Cannes, 1982.

[16] P. L. François : "Sensitivity to geometrical parameters in a dispersion free multicomponent fiber", Elec. Lett., vol.18, 11, pp.484-486, 1982.

[17] S. Kawakami and S. Nishida : "Characteristics of a doubly clad optical fiber with a low-index inner cladding", IEEE, J. Quant. Elect., vol. QE 10-12, pp.879-887, 1974. (See also vol. QE 11, pp.130-138, 1975).

[18] P. K. Mishra and I. C. Goyal : "Single-mode graded-core w-type fibers with low dispersion over a wide spectral range", Opt. Com., vol.49, pp.413-417, 15 April 1984.

[19] L. Jeunhomme, J. J. Bernard, P. Dupont, C. Lavanant, A. Tardy and M. Carrat : "Single-mode fibers for land cables : design, realisation and experimentation", Workshop on monomode techniques in the local network, Lannion, 17 April 1984.

[20] L. G. Cohen, W. L. Marmel and S. J. Jang : "Low loss quadruple clad single-mode light guides with dispersion below 2 ps/km over the 1. 28-1. 65 µm wavelength range", Elect. Lett., vol.18, pp.1023-1024, 1982.

[21] V. A. Bhagavatula and P. E. Blazyk : "Single-mode fiber with segmented core", 6th optical meeting on optical fiber communications, OSA, IEEE, MF5, pp.10-11, New Orleans, March 1983.

[22] E. M. Dianov : "Optical solitons in fibres", Europhysics News, vol.23, pp.23-26, 1992.

[23] A. Hasegawa and T. Tappert : "Transmission of stationary nonlinear optical pulses in dispersive dielectric fibers .I. Anomalous dispersion", Appl. Phys. Lett., vol.23, no.3, pp.142-149, 1973.

[24] R. H. Stolen and C. Lin : "Self-phase-modulation in silica optical fibers", Phys. Review, vol. A17, no.4, pp.1448-1453, 1978.

[25] L. F. Mollenauer, R. H. Stolen and J. P. Gordon : "Experimental observation of picosecond pulse narrowing and solitons in optical fibers", Phys. Rev. Lett., vol.45, no.13, pp.1095-1098, 1980.

[26] G. Miller : "Fiberoptics Industry Report", Laser Focus World, p.44, Pennwell Pub. USA, July 1992.

[27] L. F. Mollenauer and K. Smith : "Soliton transmission over more than 6000 km in fiber with loss periodically compensated by Raman gain", ECOC'89 Proc., vol.2, pp.71-78, 1989.

[28] A. Cozannet, J. Fleuret, H. Maître, M. Rousseau : "Optique et télécommunication", Eyrolles Ed., 1981.

[29] J. E. Midwinter : "Optical fiber for transmission", J Wiley & Sons, Chichester, 1979.

[30] K. Chida, F. Hanawa and M. Nakahara : "Fabrication of OH-free multimode fiber by vapor phase axial deposition", J. Quant. Elect., vol QE-18, no.11, pp.1883-1899, Nov. 1982.

[31] From manufacturers catalogues (for example, DWF from Corning)

[32] R. J. S. Bates, J. D. Spalink, S. J. Butterfield, J. Lipson, C. A. Barrys, J. P. Lee and R. A. Logan : "1.3 µm/1.5 µm bidirectional WDM optical fiber transmission system experiment at 144 Mbit/s", Elec. Lett., vol.19, no.13, pp.458-459, 23 June 1983.

[33] M. Poulain : "Les matériaux vitreux pour fibres optiques", Opto 84, ESI Publications, 15-17 Mai 1984.

[34] P. Delansay: "Lasers semiconducteurs", Japon Optoélectronique, no.9, p.11, March 1992.

[35] Newsbreaks, Editorial, Laser Focus World, p.13, March 1992.

[36] I. Mito, T. Torikai, M. Yamaguchi, S. Fujita, N. Henmi and K. Kobayashi : "High speed single frequency laser diodes and wide bandwidth detectors for multi Gb/s fiber optic communications systems", 10.2.1, CH 2538, 7/88/0000-0306,IEEE, pp.306-312, 1988.

[37] J. P. Laude, Ph. Gacoin, J. Flamand, J. C. Gautherin and D. Lepère : "Le multiplexage de longueurs d'onde", Opto 82 Conf. Proc., pp.144-147, Masson Ed., Paris, 1982.

[38] J. P. Laude : "Les multiplexeurs de longueurs d'onde en télécommunication optique", J. Optics, vol.15, no.6, pp.419-423, Paris, 1984.

[39] J. P. Laude , F. Bos and I. Long : "LED spectral slicing wavelength division multiplexers for single mode fiber network", Opto 91 Proc., pp.397-398, Masson Ed., 1991.

[40] M. Hirao, S. Tsuji, K. Mizuishi, A. Doi and M. Nakamura : "Long wavelength in GaAsP/InP lasers for optical fiber communications systems", J. Opt. Com., vol.1, pp.10-14, 1980.

[41] J. P. Laude : "Wavelength division multiplexers : review of some devices proposed recently", EFOC LAN 85 Proc., pp.231-238, Geneva, Information Gatekeepers Inc Boston, 1985.

[42] E. Pelletier and P. Bousquet : "Filtres optiques interférentiels pour multiplexeurs et démultiplexeurs destinés aux télécommunications par fibres", ECOC 82 Proc., Cannes, pp.532-536, published by IEE London, Sept 1982.

[43] J. Minowa and Y. Fujii : "Dielectric multilayer thin-film filters for WDM transmission systems", J. Lightwave Tech., vol.LT-1, no.1, March 1983.

[44] M. Boitel and A. Hamel : "Structures multicouches Fabry-Perot adaptées aux réseaux de vidéocommunication", Le vide et les couches minces, no.223, pp.293-7, 1984

[45] H. Trimmel, H. F. Mahlein, A. Reichelt and B. Stummer : "4 channels WDM - Transmission of 2×565 Mb/s plus 2×140 Mb/s on a single mode fiber", ECOC'84 Proc., pp.262-263, VDE-Verlag Ed Berlin, 1984.

[46] M. O. Kuno and S. Kobayashi : "Low loss optical multi/demultiplexer compatible with SM fibers", Trans. IEICE, vol. E72, no.6, pp.736-741, June 1989.

[47] S. Sugimoto *et al* : "Wavelength division two way fibre. Optic transmission experiments using microoptic duplexer", Elec. Lett., vol.14, no.1, pp.15-17, Jan.1978.

[48] K. Sano, Y. Fujii and J. Minowa :"Low-loss optical multi/demultiplexer for subscriber loop system", Review of electrical communication laboratories, vol.32, no.4, pp.608-618, 1984.

[49] B. Hillerich, M. Rode and E. Weidel : "Duplexer with hybrid integrated light emitter and detector", ECOC' 84 Proc., pp.166-167, Stuttgart, VDE Verlag Berlin Ed., 1984.

[50] P. Gacoin, J. C. Gautherin, F. Bos, D. Lepère, J. Flamand and J. P. Laude : "Un multiplexeur trois voies à faibles pertes", Revue Physique Appliquée, vol.19, pp.99-109, 1984.

[51] K. Nosu, H. Hashimoto and K. Hashimoto : "Projects in opto communications", IEE reprints, series 3, pp.183-184, 1980.

[52] Sh. Ishikawa, K. Takahashi and K. Dor : "Multireflection wavelength-division multiplexer/demultiplexer with the wavelength tunability", 6th ECOC IEE, no.190, pp.298-301, Sept.1980.

[53] R. Reichelt, H. F. Mahlein and G. Winzer : "Low-loss lensless wavelength -division multiplexers [using beam splitter principle]", 6th ECOC IEE, no.190, pp.294-297, Sept. 1980.

[54] H. F. Mahlein , H. Michel, W. R. Rauscher, A. Reichelt and G. Winzer :"Interference filter all-fibre directional coupler for WDM", Elec. Lett., vol.16, no.15, pp.584-5, 1980.

[55] Tamura *et al* : "Multiplexeur/démultiplexeur hybride optique de longueurs d'onde", Europ. Patent 0153722, 1984.

[56] M. C. Hutley : "Diffraction gratings", Acad. Press, London, 1982.

[57] R. Petit : "Electromagnetic theory of gratings", Springer-Verlag, Berlin, 1980.

[58] J. P. Laude and J. Flamand : "Herstellung von Bengungsgitten für Spektrometrie und Optoelektronik", Feinwerktechnik und Messtechnick 94 (1986) 5, Carl Hanser Verlag, München, 1986.

[59] J. M. Lerner, J. Flamand, J. P. Laude, G. Passereau and A. Thévenon : "Diffraction gratings ruled and holographic - A review", SPIE, vol.240-14, pp.82-88, 1980.

[60] As above : "Aberration corrected holographically recorded diffraction gratings", SPIE, vol.240-13, pp.72-81, 1980.

[61] J. P. Laude, brevet européen 332790, date de priorité : 18 mars 1988, et M. Nevière, D. Maystre and J. P. Laude : "Perfect blazing for transmission gratings", J. Opt. Soc. Am. A, vol.7, no.9, pp.1736-1739, Sept.1990.

[62] M. Detaille, M. Duban and J. P. Laude : "Réalisation de réseaux de diffraction pour le satellite astronomique D2B", Perspectives 91, Jobin Yvon documentation, Feb. 1976.

[63] D. Maystre, J. P. Laude, P. Gacoin, D. Lepère and J. P. Priou : "Gratings for tuneable lasers : using multidielectric coatings to improve their efficiency", Appl. Opt., vol.19, no.18, pp.3099-3102, 15 Sept.1980.

[64] J. P. Laude, J. Flamand, J. C. Gautherin, D. Lepère, F. Bos, P. Gacoin, A. Hamel : "Multiplexeur de longueurs d'onde à micro-optique pour fibres unimodales", Sixièmes journées nationales d'optique guidée, Issy, 20-21 mars 1985.

[65] K. Aoyame and J. Minowa : "Low loss optical demultiplexer for WDM systems in the 0.8 μm wavelength region", Appl. Opt., vol.18, no.16/15, pp.2854-2836, 15 August 1979.

[66] R. Watanabe, K. Nosu and Y. Fujii : "Optical grating multiplexer in the 1.1-1.5 μm wavelength region", Elec. Lett., vol.16, no.3, p.107, 1980.

[67] W. J. Tomlinson : "Wavelength multiplexing in multimode optical fibers", Appl. Opt., vol.16, no.8, pp.2180-2194, August 1977.

[68] B. D. Metcalf and J. F. Providakes : "High-capacity wavelength demultiplexer with a large-diameter Grin rod lens", Appl. Opt., vol.21, no.5, pp.794-796, March 1982.

[69] M. Seki, K. Kobayashi, Y. Odagiri, M. Shikad, T. Tanigawa and R. Ishikawa: "20-channel micro-optic grating demultiplexer for 1.1-1.6 μm band using a small focusing parameter graded-index rod lens", Elec. Lett., vol.18, no.6, pp.257-258, 18 March 1982.

[70] W. J. Tomlinson : "Wavelength division multiplexer", US patent, Doc. 4, III, 524, Sept. 1978.

[71] Hans, G. Finke, A. Nicia and D. Rittick : "Optische Kugellinsen demultiplexer", Optische Nachrichtentechnik, Ntz Bd 37, pp.346-351, 1984.

[72] J. P. Laude : "High brightness monochromator using optical fibers", Conférence Opto, Proc., p.60, ESI Ed.Paris, 1980.

[73] J. P. Laude and J. Flamand : "Un multiplexeur-démultiplexeur de longueur d'onde (configuration Stimax)", Revue Opto no.3, pp.33-34, Février 1981.

[74] L. Mannschke : "Microcomputer aided design and realisation of low insertion loss wavelength multiplexer and demultiplexer", SPIE Proc., vol.399, pp.92-97, 1983.

[75] L. Mannschke : "A multiplexer/coupler with tapered graded-index glass fibers and a grin rod lens", ECOC'84, Proc., pp.164-165, IEE Ed., 1984.

[76] J. P. Laude, J. Flamand, J. C. Gautherin, D. Lepère, P. Gacoin, F. Bos : "STIMAX, a grating multiplexer for monomode or multimode fibers", ECOC'83, Proc., p.417, Geneva, Elsevier Science Pub., 1983.

[77] R. Watanabe, K. Nosu, T. Harada and T. Kita : "Optical demultiplexer using a concave grating in then 0.7-0.9 μm wavelength region", Elec. Lett., vol.16, no.3, p.106, 1980.

[78] Y. Fujii and J. Minowa : "Cylindrical concave grating utilising thin silicon chip", Elec. Lett., vol.17, no.24, pp.934-936, 26 Nov.1981.

[79] T. Kita and T. Harada : "Use of aberration corrected concave grating in optical demultiplexing", Appl. Opt., vol.22, no.6, pp.819-825, 15 March 1983.

[80] A. M. Koonen and A. Wismeijer: "Optical devices for wavelength division multiplexing systems", Philips Tel. Rev., vol.40, no.2, pp.102-110, July 1982.

[81] R. Watanabe, Y. Fujii, K. Nosu and J. I. Minowa : "Optical multi/demultiplexer for single-mode fibers transmission", IEEE J. Quant. Elect., vol.QE 17, no.6, pp.974-981, June 1981.

[82] A. Leboutet, R. Auffret, G. Claveau, M. Guibert, J. Moalic, L. Pophillat, Y. Sorel and A. Tromeur : "Wavelength division multiplexing in the 1.5 µm window : an installed link", Elec. Lett., vol.20, pp. 834-835, 27 Sept.1984.

[83] J. Hegarty, S. D. Poulsen, K. A. Tackson and P. Raminov : "Low-loss single-mode wavelength division multiplexer with etched fibre array", Elec. Lett., vol.20, no.17, pp.685-686, 16 August 1984.

[84] J. P. Laude :"Les multiplexeurs monomodes à réseaux de diffraction. Largeurs comparées des fonctions de transmission spectrales de composants de 2 à 20 voies", Opto'90, Paris Actes, pp.480-480, ESI Ed., 1990.

[85] J. P. Laude, F. Bos, D. Fessard, J. Flamand, P. Gacoin, J.C. Gautherin, D. Lepère and I. Long : "Results obtained with 12 multiplexers using single-mode fibers", EFOC LAN 90 Proc., pp.156-159, Information Gatekeeper Inc. Ed., 1990.

[86] D. R. Wisely : "High performance 32 channels HDWDM multiplexer with 1 nm channel spacing and 0.7 nm bandwidth", SPIE, vol.1578, pp.170-176, 1991.

[87] E. Snitzer, Advances in Quantum Electronics, JR Singer Ed., Columbia University Press, NY, p.348, 1961.

[88] T. Ozeki and B. S. Kawasaki : "Optical directional couplers using tapered sections in multimode fibers", Appl. Phys. Lett., vol.28, p.528, 1976.

[89] M. K. Barnoski and H. R. Friedrich : "Fabrication of an access coupler with single-strand multimode fiber waveguide", Appl. Opt., vol.15, no.11, p.2629, 1976.

[90] T. Ozeki and B. S. Kawasaki : "Mode behavior in a tapered multimode fiber", Elec. Lett., vol.12, no.6, pp.407-408, 5th August 1976.

[91] Y. Tsujimoto, H. Serizawa, K. Hattori and M. Fukai : "Fabrication of low-loss 3 dB couplers with multimode optical fibres", Elec. Lett., vol.14, pp.157-158, 1978.

[92] R. A. Bergh, G. Kotter and H. J. Shaw : "Single mode fibre optic directional coupler", Elec. Lett., vol.16, pp.260-261, 27 March 1980.

[93] O. Parriaux, S. Gidon and A. A. Kuznetsov : "Distributed coupling on polished single mode optical fibres", Appl. Opt., vol.20, no.14, 15 July 1981.

[94] M. Digonnet and H. J. Shaw : "Wavelength multiplexing in single-mode fibre couplers", Appl. Opt., vol.22, no.3, pp.484-491, 1 February 1983 and J. Quant. Elect., vol. QE18, no.746, 1982.

[95] D. Marcusse : "Directional coupler with filter using dissimilar optical fibres", Elec. Lett., vol.21, pp.26-27, 1985.

[96] O. G. Leminger and R. Zengerle : "Determination of single mode fiber coupler design parameters from loss measurements", J. Lightwave Tech., vol. JLT3, no.4, p.864, August 1985.

[97] M. S. Whalen and K. L. Walker : "In-line optical fibre filter for wavelength multiplexing", Elect. Lett., vol.21, no.17, pp.724-725, 15 August 1985.

[98] S. K. Sheem and T. G. Giallorenzi : "Single-mode fiber-optical power divider, encapsulated etching technique", Opt. Lett., vol.4, no.29, 1979.

[99] P. J. Severin and A. P. Severijns : "Passive components for multimode fibre optics networks", IOOC ECOC'85 Proc., pp.453-6, 1985.

[100] D. Kreit and R. C. Youngquist : "Report on polished single mode fibre couplers", Race project RDP no.2031, Commission of European communities, Brussels, April 1986.

[101] G. Georgiou and A. C. Boucouvalas : "High isolation single mode wavelength division multiplexer/demultiplexer", Elec. Lett., vol.22, no.2, pp.62-63, 16 Jan.1986.

[102] P. Kopera, M. Corke, K. Sweeney and M. Keur : "Passive single mode fibre components", 36th electronic components conference, Proc. 1986, no.86, CH 2302-8, 1986.

[103] K. Fussgaenger, W. Koester, H. D. Saller and T. Vollmer : "$4\lambda \times 560$ Mbit/s WDM system using 3 wavelengths selective fused single-mode fibre couplers as multiplexer", ECOC' 86 Proc., pp.447-450, 1986.

[104] M. S. Yataki, D. N. Payne and M. P. Varnham : "All fibre wavelength filters using concatenated fused taper couplers", Elec. Lett., vol.21, no.6, pp.248-249, 14 March 1985.

[105] K. W. Fussgaenger and R. H. Rossberg : "Uni and bi-directional $4\lambda \times 560$ Mb/s. Transmission systems using WDM devices based on wavelength selective fused single-mode fibre couplers", IEEE Journal on selected area in Com., vol.8, no.6, pp.1032-1042, August 1990.

[106] X. B. Luo, L. Hu and W. G. Lin : "Very high isolation all fiber wavelength division multi-demultiplexer", Microwave and optical technology letters, vol.3, no.4, pp.116-117, April 1990.

Bibliography

[107] G. Winzer, W. Döldisen, C. Cremer, F. Fiedler, G. Heise, R. Kaiser, R. März, H. H. Mahlein, L. Mörl, H.P. Nolting, W. Rehbein, M. Schienle, G. Schulte-Roth, G. Unterbörsch, H. Unzeitig and U. Wolff : "Monolithically integrated detector chip for a two-channel unidirectional WDM link at 1.5 µm", IEEE journal on selected areas in com., vol.8, no.6, August 1990.

[108] J. P. Laude : "Wavelength division multiplexers technological trends", EFOC LAN 87 Proc., pp.85-90, Information Gatekeepers Inc USA, June 1987.

[109] Editorial, interview of M. Nakamura : "Six lasers on a chip with waveguides give multiplexing", Electronics, 3E/4E, 2 February 1978.

[110] T. P. Lee, C. A. Burrus and A. G. Dentai : "Dual wavelength surface emitting in GaAsP LED", Elec. Lett., vol.16, pp. 841-845, 1980.

[111] N. Bouadma, J. C. Bouley, J. Riou : "Dual wavelength (GaAl)As lasers", Elect. Lett., vol.18, no.20, p.871, Sept.1982.

[112] J. P. Van der Ziel, H. Temlin and R. A. Logan : "Wavelength multiplexing of 1. 31 µm in GaAsP buried crescent laser arrays", Appl. Phys. Lett. (USA), vol.43, no.5, pp.401-403, 1983.

[113] A. K. Chin, B. H. Chin, I. Cambinel, C. L. Zipfel and G. Minneci : "Practical dual-wavelength ligth-emitting double diode", J. Appl. Phys., vol.57, no.12, pp.5519-5522, 15 June 1985.

[114] L. A. Koszi and N. A. Olsson : "Wavelength division multiplexed optical transmission system with single mode fibre and dual wavelength hybrid laser light source", Elec. Lett., vol.22, no.21, pp.1102-1103, 9 October 1986.

[115] N. K. Dutta, T. Cella, J. L. Zilko, D. A. Ackerman, A. B. Piccirilli and L. I. Greene : "InGaAsP closely spaced dual wavelength laser", Cleo Th 16, Baltimore, April-May 1987.

[116] Masahi Nakao : "Development and application of 20 wavelength DFB laser array by SOR lithography", JEE, pp.32-35, May 1989.

[117] T. Suhara, Y. Handa, H. Nishihara and J. Koyama : "Monolithic integrated microgratings and photodiodes for wavelength demultiplexing", Appl. Phys. Lett., vol. 40, no.2, pp.120-122, 15 Jan 1982.

[118] A. Larsson, P. A. Andrekson, P. Andersson, J. Salzman and A. Yariv : "High speed dual wavelength demultiplexing and detection in a monolithic superlattice P-I-N waveguide detector array", Appl. Phys. Lett., vol.49 (5), pp.233-235, 4 August 1986.

[119] T. Miyazawa, T. Tagawa, H. Iwamura, O. Mikami and M. Naganuma : "Two-wavelength demultiplexing pin GaAs/AlAs photodetector using partially disordered multiple quantum well structures", Appl. Phys. Lett., vol.55, no.9, pp.828-829, August 1989.

[120]   M. S. Ünlü, K. Kishino, J. Y. Reed, L. Arsenault and H. Morkoç : "Wavelength demultiplexing heterojunction phototransistors", Elec. Lett., vol.26, no.22, pp.1857-1858, 25 October 1990.

[121]   M. S. Ünlü, A. L. Demirel, S. Strite, S. Tasiran, A. Salvador and H. Morkoç : "Wavelength demultiplexing optical switch", Appl. Phys. Lett., vol.60 (15), pp.1797-1799, 13 April 1992.

122]   C. Bornholdt, D. Trommer, G. Unterbörsch, H. G. Bach, F. Kappe, W. Passenberg, W. Rehbein, F. Reier, C. Schramm, R. Stenzel, A. Umbach, H. Venhaus and C. M. Weinert : "Integrated demultiplexer-receiver on InP", Appl. Phys. Lett., vol.60, no.8, pp.971-973, 24 February 1992.

[123]   R. C. Alferness : "Guided wave devices for optical communication", IEEE Journal of QE, vol. QE17, no.6, pp.946-958, June 1981.

[124]   R. Regener and W. Sohder : "Loss in low-finesse Ti:NbO$_3$ optical waveguide resonators", Appl. Phys. B, vol.36, no.3, pp.143-147, March 1985.

[125]   R. C. Alferness, L. L. Buhl and M. D. Divino : "Low loss fibre coupled waveguide directional coupler modulator", Elec. Lett., vol.18, pp.490-491, 1982.

[126]   M. Fukuma, J. Noda and H. Iwasaki : "Optical properties in titanium-diffused LiNbO$_3$ strip waveguides", J. Appl. Phys., vol.49, no.7, pp.3693-3698, 1978.

[127]   R. C. Alferness, S. K. Korotky, L. L. Buhl and M. D. Divino : "High-speed low-loss low-drive-power traveling-wave optical modulator for $\lambda = 1.32\mu m$", Elec. Lett., vol.20, pp.554-555, 1984.

[128]   R. C. Alferness and L. L. Buhl : "Low loss, wavelength tunable, waveguide electro-optic polarization controller for $\lambda = 1.32\ \mu m$", Appl. Phys. Lett., vol.47, p.1137, 1985.

[129]   O. Mikanni : "LiNbO$_3$ coupled-waveguide TE/TM mode splitter", Appl. Phys. Lett., vol.36, no.7, pp.491-493, 1980.

[130]   R.C. Alferness and J.J. Veselka : "Simultaneous modulation and wavelength multiplexing with a tunable Ti : Li Nb O$_3$ directional coupler filter", Elec. Lett., vol.21, no.11, pp.466-467, 23 May 1985.

[131]   H. A. Haus and N. A. Whitaker Jr : "Elimination of crosstalk in optical directional couplers", Appl. Phys. Letters, vol.46, no.1, pp.1-3, 1 Jan 1985.

[132]   J. P. Lin , R. Hsiao and S. Thaniyavarn : "Four-channel wavelength division multiplexer on Ti : Li Nb O$_3$", Elec. Lett., vol.25, no.23, pp.1608-1609, 9 November 1986.

[133]   M. De Sario, A. D'Orazio, V. Lanave and V. Petruzzelli : "Integrated optical demultiplexer for WDM systems", Dep di Elettrotechnica ed Elettronica, Via Re David 200, 0125 Bari, Italy, work under PF-Madess of Italian Council of Research (CNR), 1988.

[134] K. S. Chiang : "Dual effective index method for the analysis of rectangular dielectric waveguides", Appl. Opt., vol.25, no.13, pp.2169-2174, 1986.

[135] B. K. Garside and P. E. Jessop : "New semi conductor materials and structures for electro optical devices", Can. J. Phys., vol.63, p.801, 1985.

[136] Belekolov, E. M. Dianov and A. A. Kusnetzov : "Optical demultiplexer with a glass slide", EOOC 81, Tech. D., p.66, 1981.

[137] Y. Fujii, J. Minowa and Y. Yamada : "Optical demultiplexer utilizing an Ebert mounting silicon grating", Journal of lightwave tech., vol. LT2, no.5, pp.731-734, October 1984.

[138] M. Masson, H. Royer, R. Dupeyrat, J. P. Laude : "Méthode d'exitation du spectre Raman d'une couche mince utilisant un réseau à grand nombre de traits", Opt. Com., vol.20, no.3, pp.443-445, March 1977.

[139] T. Von Lingelsheim : "Planar optical demultiplexer with chirped grating for WDM fibre systems", IEEE Proceedings, vol.131, no.5, pp.290-294, Oct. 1984.

[140] T. Izawa and H. Nakagome : "Optical waveguide formed by electrically induced migration of ions in glass plates", Appl. Phys. Lett., vol.21, no.12, pp.584-586, 1972.

[141] G. H. Chartier, P. Jaussand, A. D. de Oliveira, O. Parriaux : "Optical waveguides fabricated by electric-field controlled ion exchange in glass", Elec. Lett., vol.14, no.5, pp.132-134, March 1978.

[142] R. Walker and C. D. W. Wilkinson : "Integrated optical waveguiding structures made by silver ion-exchange in glass: directional coupler and bends", Appl. Opt., vol.22, pp.1929-1936, 1983.

[143] F. M. Ernsberger : "Lectures on glass and technology", Rensselaer Polytech. Inst. Troy, N.Y., Glass Industry., vol.47, pp.542-545, October 1966.

[144] T. Findakly : "Glass waveguides by ion exchange : a review", Optical Engineering, vol.24, no.2, pp.244-250, March-April 1985.

[145] Y. Okamura, S. Yoshinaka and S. Yamoto, Appl. Opt., vol.22, p.3892, 1983.

[146] M. Lofti Gomoa, G. Chartier : "Etude expérimentale du multiplexage optique en longueur d'onde à l'aide de guides planaires monomodaux", CR Acad. Sc. de Paris, T. 301, série II, no.11, pp.775-778, 1985.

[147] a) K. W. Murphy : "An integrated optics technology for the production of Photocor TM fiber-optics components", Corning technical report, TR-70, Oct. 1987.

[147] b) C. Nissim, A. Béguin, P. Laborde, D. Lerminiaux and M. McCourt : "Low loss SM wavelength division multiplexers fabricated by ion exchange in glass", EFOC 90 Proc., pp.114-117, Munich, 1990.

[148] Imoto, Katsuyuki : "Waveguide type optical multiplexer/demultiplexer", European patent no.0 306 956, A2, 08.09.88.

[149] E. Okuda, I. Tanaka, T. Yamasaki : "Planar gradient-index glass waveguide and its applications to a 4 port branched circuit and star coupler", Appl. Opt., vol.23, no.11, pp.1745-1748, June 1984.

[150] N. Goto and G. L. Yip: "Y-junction wavelength multi/demultiplexer by K^+ ion exchange in glass for 1.3 and 1.5 µm", SPIE, vol.1141, 5th European conference on integrated optics, ICIO'89, 1989.

[151] N. Goto and G. L. Yip : "Y-branch wavelength multi/demultiplexer for 1.30 and 1.55 µm", Elec. Lett., vol.26, no.2, pp.102-103, 18 Jan 1990.

[152] M. Seki, R. Sugawara and Y. Hanada : "Novel design for a high performance guided-wave multi/demultiplexer in glass", J. Modern Optics, vol.36, no.6, pp.797-808, 1989

[153] T. Negami, H. Haga and S. Yamamoto : "Guided wave optical wavelength demultiplexer using an asymetric Y junction", Appl. Phys. Lett., vol.54, no.12, pp.1080-1082, 20 March 1989.

[154] S. Suzuki, S. Sekine, K. Shuto, Y. Ueoka and I. Nishi : "High Δ glass waveguide multi/demultiplexers with small device size and low wavelength response sensitivity", Trans. IEICE, vol. E73, no.1, Jan. 1990.

[155] P. Barlier, C. Nissim and L. Dohan : "Passive integrated optics components for fiber optics communication in moldable glass", IOOC/ECOC 85 Proc., pp.187-190, Venise, October 1985.

[156] A. Tervonen, P. Pöyhönen, S. Houhanen and M. Tahkokorpi : "Channel waveguide Mach-Zehnder interferometer for wavelength splitting and combining", SPIE, vol.1513, pp.71-75, 1991.

[157] N. Kuzuta and E. Hasegawa : "An optical demultiplexer using a flexible replica grating, an embedded optical waveguide and an uneven fiber space array", J. Appl. Phys., vol.64, no.7, pp.3745-3748, 1 October 1988

[158] Masayuki Takami : "Wavelength division system transmits five wavelengths by integrated optics technology", JEE (Japan), vol. 23, no.74.7, p.99, 1986.

[159] G. Grand, S. Valette, G. J. Cannell, J. Aarnio and M. Del Giudice : "Fibre pigtailed silicon based low cost passive optical components", ECOC'90 Proc., pp.525-528, Amsterdam, September 1990.

[160] G. Grand, B. Corselle, J. P. Jadot, E. Parrens, A. Fournier, A. M. Grouillet and S. Valette : "16-channel optical wavelength multiplexer/demultiplexer integrated on silicon substrate", EFOC'LAN 91 Proc., pp.264-267, London, June 1991.

[161] Nasa Tech Brief, "Integrated grating spectrometer", April 1990.

[162] P. S. Henry : "Optical wavelength division multiplex", CH 2827 - 4/90/0000 - 1508, IEEE, 1990.

[163] A. J. N. Houghton, D. A. Andrews, G. J. Davies and S. T. D. Ritchie : " Low loss optical waveguides in MBE-grown GaAs/GaAlAs heterostructures", Opt. Com., vol.46, p.164, 1983.

[164] D. A. Andrews, E. G. Scott, A. J. N. Houghton, P. M. Rodgers and G. T. Davies : "The growth of GaAlAs/GaAs guide wave devices by molecular epitaxy", J. Vac. Sci. Technol., vol.B3, no.3, pp.813-815, May-June 1985

[165] C. De Bernardi, S. Morasca, C. Rigo, B. Sordo, A. Stano, I.R. Croston and T.P. Young : "Wavelength demultiplexer integrated on AlGaAs/InP for 1.5 µm operation", Elec. Lett., vol.25, no.22, pp.1488-1489, 26th October 1989.

[166] I. R. Croston and T. P. Young :"Design of an InGaAlAs/InP '3 mi' wavelength division demultiplexer employing a novel mode transformer", Elec. Lett., vol.26, no.5, pp.336-337, 1990.

[167] C. Bornholdt, F. Kappe, M. Nolting, R. Stenzel, H. Venghaus and C. M. Weinert : "WDM-Bauelemente auf der basis von InGaAsP/InP", Richtkopplern ITG Fachber (Germany), vol.112, pp.119-124, 1990.

[168] C. Cremer : "Integriert optischer Spektrograph für WDM Komponenten", ITG Fachber (Germany), vol.112, pp.125-130, 1990.

[169] J. B. D. Soole, A. Scherer, H. P. Leblanc, N. C. Andreakis, R. Bhat and M. A. Koza : "Monolithic InP based grating spectrometer for wavelength division multiplexed system at 1.5 µm", Elec. Lett. (UK), vol.27, no.2, pp.132-134, 17 Jan 1991.

[170] C. M. Ragdale, T. J. Reid, D. C. J. Reid, A. C. Carter and P. J. Williams : "Integrated laser and add-drop optical multiplexer for narrowband wavelength division multiplexing", Elec. Lett., vol.28, no.8, pp.712-714, April 1992.

[171] a) A. H. Gnauck et al : Paper PD 26, Optical Fiber Communication Conference, San Francisco, Jan. 1990.

[171] b) S. Sekine , K. Sato, F. Kano, Y. Kondo and Okamoto : " A four channels,wavelength tuneable MQW-DFB laser array module for optical FDM transmission systems", ECOC'92 Proc., vol.1, pp.173-176, 1992.

[172] M. R. Wang, R. T. Chen, G. J. Sonek, T. Jannson and H. T. Lu : "Wavelength division demultiplexing in the infrared using holographically processed polymer microstructure waveguides", SPIE, vol.1347, pp.560-565, 1990.

[173] T. Suhara, H. Nishihara and J. Koyama : "High performance focusing grating coupler fabricated by electron beam writing", Topical meeting on integrated and guided-wave optics, Kissimmee, Florida, Paper Th D4-1, April 1984.

[174] G. N. Lawrence and P. J. Cronkite : "Physical optics analysis of the focusing grating coupler", Appl. Opt., vol.27, no.4, p.672, 15 Feb 1988.

[175] P. J. Cronkite and G. N. Lawrence : "Focusing grating coupler design method using holographic optical elements", Appl. Opt., vol.27, no.4, p.679, 15 Feb 1988.

[176] S. Ura, M. Morisawa, T. Suhara and H. Nishihara : "Integrated optic wavelength demultiplexer using a coplanar grating lens", Appl. Opt., vol.29, no.9, p.1369, 20 March 1990.

[177] D. A. Bryan, C. R. Chubb, J. K. Powers, W. R. Reed : "Integral grating coupler on an optical fibre", SPIE, vol.574, pp.56-61, 1985.

[178] C. M. Ragdale, D. Reid, D. J. Robbins, J. Buns and I. Bennion : "Narrowband fiber grating filters", IEEE Journal on selected areas in communications, vol.8, no.6, pp.1146-1150, August 1990.

[179] H. Takahashi, S. Suzuki, K. Kato and I. Nishi : "Arrayed waveguide grating for wavelength division multi/demultiplexer with nanometer resolution", Elec. Lett., vol.26, no.2, pp.87-88, 18 January 1990.

[180] H. Takahashi, Y. Hibino and Y. Nishi : "Polarization insensitive arrayed waveguide grating wavelength multiplexer on silicon", Opt. Lett., vol.17, no.7, pp.499-501, April 1, 1992.

[181] J. P. Laude and J. Lerner : "Wavelength division multiplexing/demultiplexing (WDM) using diffraction gratings", Proc. SPIE 503, pp.22-28, San Diego, 1984.

[182] M. H. Reeve, A. R. Hunwicks, W. Zhao, S. G. Mettiley, L. Bickers and S. Hornung : "LED spectral slicing for single mode local loop applications", Elec. Lett., vol.24, no.7, pp.389-390, 31 March 1988.

[183] A. R. Hunwicks, L. Bickers and P. Rogerson : "A spectrally sliced single-mode optical transmission system installed in the UK local loop networks", Globecom 89, IEEE Global Telecommunication Conference and Exhibition, vol.3, pp.1303-1307 (Pub.IEEE 1989, New York), Dallas, 27-30 Nov 1989.

[184] S. S. Wagner and T. E. Chapuran : "Broadband high density WDM transmission using superluminescent diodes", Elec. Lett. (UK), vol.26, no.11, pp.696-697, 24 May 1990.

[185] A. R. Hunwicks : "The effect of transmitter wavelength variations in spectrally sliced optical transmission systems", SPIE, vol.1179, pp.15-25, Fiber networking and Telecommunication, 1989.

[186] L. Bersiner, D. Rund : "Bi-directional WDM transmission with spectrum sliced LEDs", Opt. Com., vol.11-2, pp.56-59, 1990.

[187] JP Laude : "Multiplexeurs multiples à réseau unique", Photon 1983, SPIE, vol.403, pp.253-260, 1984.

[188] B. R. Clarke and M. Bornhoft : "The performance of Fabry-Perot interferometers as wavelength selective devices for optical transmission systems", ATR, vol.22, no.2, pp.9-22, 1988.

[189] A. A. M. Saleh and J. Stone : "Two-stage Fabry-Perot filters as demultiplexers in optical FDMA LAN's", J. Lightwave Tech., vol.7, no.2, pp.323-330, Feb.1989 .

[190] A. Frenkel and C. Lin : "Angle-tuned etalon filters for optical channel selection in high density wavelength division multiplexed systems", J. Lightwave Tech., vol.7, no.4, pp.615-624, April 1989.

[191] D. McMahon, W. A. Dyes and W. C. Robinson : "Novel bulk optic multichannel resonator device for close packed wavelength multiplexing", Appl. Opt., vol.28, no.13, pp.2529-2737, 1 July 1989.

[192] J. S. Harper, P. A. Rosher, S. Fenning and S. R. Mallison : "Application of miniature micromachined Fabry Perot interferometer to optical fiber WDM systems", Elec. Lett., vol.25, no.16, pp.1065-1066, 3 August 1989.

[193] K. Y. Eng, M. A. Santoro, J. Stone and T. L. Koch : "Optical FDM switch experiments with tunable fiber Fabry Perot filters", CH 2827 - 4/90/0000 - 1529, IEEE, 1990.

[194] C. M. Miller : "Characteristics and applications of high performance, tunable, fiber Fabry Perot filters", Proc. 41st Electronic conf. and Tech. conf., Atlanta, IEEE 1991, pp.489-492, May 1991.

[195] N. Goto and Y. Miyazaki : "Integrated multi-demultiplexer using acoustooptic effect for multiwavelength optical communications", IEEE Journal on selected areas in com., vol.8, no.6, pp.1160-1168, August 1990.

[196] K. W. Cheung, D. A. Smith, J. E. Baran and B. L. Heffner : "Multiple channel operation of integrated acousto optic tunable filter", Elec. Lett., vol.25, no.6, pp.375-376, 16 March 1989.

[197] K. W. Cheung, S. C. Liew, C. N. Lo, D. A. Smith, J. E. Baran and J. J. Johnson : "Simultaneous five wavelength filtering at 2.2 nm wavelength separation using integrated acoustooptic tunable filter with subcarrier detection", Elec. Lett., vol.25, no.10, pp.636-637, 11 May 1989.

[198] T. Kinoshita and K. I. Sano : "Design and performance of a tunable optical demultiplexer using an acoustooptic light deflector", Elec. and Com. in Japan, vol.72, part 2, no.6, pp.14-22, 1989.

[199] D. A. Smith, J. E. Baran, J. J. Johnson and K. W. Cheung : "Integrated optic acoustically tunable filters for WDM networks", IEEE Journal on selected areas in com., vol.8, no.6, August 1990.

[200] B. L. Heffner, D. A. Smith, J. E. Baran, A. Yi-Yan and K. W. Cheung : "Integrated optic acoustically tunable infrared optical filter", Elec. Lett., vol.24, no.25, pp.1562-1563, 8th December 1988.

[201] S. E. Harris and R. W. Wallace : "Acoustooptic tunable filter", J. Opt. Soc. Am., vol.59, no.6, pp.744-747, 1969.

[202] Y. Shimazu and S. Kikuchi : "Time and wavelength division optical distribution system using acoustooptic tunable filter", SPIE, vol.1179, Fiber networking and telecommunications, pp.34-42, 1989.

[203]   S. C. Liew : "A multiwavelength optical switch based on the acousto-optic tunable filter", SPIE, vol.1363, pp.57-61, 1990.

[204]   K. W. Cheung, D. A. Smith, J. E. Baran and J. J. Johnson, IEEE - CH 2827 - 4/90/0000-1541, 1990.

[205]   H. Heidrich, D. Hoffmann and R. I. MacDonald : "Polarisation and wavelength multiplexed bidirectional single fiber subscriber loop", Opt. Com., vol.4, pp.136-138, 1986.

[206]   J. Charlier, B. Laurent, J. Lorsignol, P. Berlioz and JL. Perbos : "Multi/demultiplexer and spectral isolator for optical inter satellites communications", SPIE, vol.1131, Optical space communications, pp.54-62, 1989.

[207]   C. F. Buhrer : European patent 0288769, "Multichannel wavelength multi/demultiplexer", 1988.

[208]   Y. Fujii and J. Minowa : "Four channel wavelength multiplexing composed of phase plates and polarizing beam splitters", Appl. Opt., vol.28, no.7, pp.1305-1308, 1 April 1989.

[209]   M. Koga, J. Minowa and T. Matsumoto : "Multi-demultiplexer using a 4 port optical circulator and interference filters", Trans IEICE, vol.E72, no.10, pp.1089-1091, October 1989.

[210]   C. Herard and A. Lacourt : "New multiplexing technique using polarization of light", Appl. Opt., vol.30, no.2, pp.222-231, 10 Jan 1991.

[211]   D. E. Rumelhart, G. E. Hinton and R. J. Williams : "Parallel distributed processing: explorations in the microstructure of cognition", Bradford Books/MIT Press, Cambridge MA, 1986.

[212]   Observatoire français des techniques avancées : "L'électronique moléculaire", Masson Ed., 1988.

[213]   M. Koga and T. Matsumoto : "A novel optical WDM demultiplexer consisting of a simple optical multimode guide and an electrical neural network", IEEE Phot. Tech. Lett., vol.2, no.7, pp.487-489, July 1990.

[214]   T. Kinoshita and K. Sano : "A programmable optical demultiplexer using a 32 element photodiode array", Electronics and Com. in Japan, vol.72, part 2, no.3, pp.71-80, 1989.

[215]   G. J. Cannell, A. Robertson, R. Worthington : "Practical realization of a high density diode coupled wavelength demultiplexer", IEEE, Journal on Selected Areas in Com., vol.8, no.6, pp.1141-1145, August 1990.

[216]   W. S. Lee, D. A. H. Spear, A. D. Smith, S. A. Wheeler and S. W. Bland : "Monolithic eight-channel photoreceiver array OEICs for HDWDM applications at 1.55 µm", Elec. Lett., vol.28, no.7, pp.612-614, 1992.

[217] E. E. Harstead, J. Ocenasek : "Short wavelength transmission over single mode fiber optimized for long wavelengths", SPIE, vol.1578, pp.86-93, 1991.

[218] J. C. Simon : "Semi-conductor laser amplifier for SM optical fiber communications", J. Opt. Com., vol.4, pp.51-62, January 1983.

[219] J. C. Simon, I. Joindot, J. Charil and D. Huibon Hoa : IOOC'81, San Francisco, 27-29 April 1981.

[220] T. Mukai and Y. Yamamoto : "Gain, frequency bandwidth and saturation output power of AlGaAs DH laser amplifiers", IEEE, J. Quant. Elect., vol.QE 17, no.6, pp.1028-1034, 1981.

[221] L. Kazovsky and J. Werner : "Multichannel optical communications using tunable Fabry Perot amplifiers", Appl. Opt., vol.28, no.3, pp.553-555, 1 February 1989.

[222] R. P. Braun, R. Ludwig and R. Molt : "Ten-channel coherent optic fibre transmission using an optical traveling wave amplifier", ECOC'86, Tech. Dig., vol.III, pp.29-32, September 1986.

[223] M. C. Oberg, N. A. Olsson, L. A. Zoszi and G. Przybylek : "313 km transmission experiment at 1 Gb/s using optical amplifiers and a low chirp laser", Elec. Lett., vol.24, pp.38-39, 1988.

[224] G. Coquin, H. Kobrinski, C. E. Zah, F. K. Shokoohi, C. Caneau and S. G. Menocal : "Simultaneous amplification of 20 channels in a multiwavelength distribution system", IEEE Phot. Tech. Lett., vol.1, no.7, pp.176-178, July 1989.

[225] M. Koga and T. Matsumoto : "The performance of a traveling wave type semi-conductor laser amplifier as a booster in multiwavelength simultaneous amplification", J. Lightwave Tech., vol.8, no.1, pp.105-113, January 1990.

[226] A. R. Chraplyvy and R. W. Tkach : "Narrowband tunable optical filter for channel selection in densely packed WDM systems", Elec. Lett., vol.22, pp.1084-1085, 1986.

[227] R. W. Tkach, A. R. Chraplyvy and R. M. Derosier : "Performance of a WDM network based on stimulated Brillouin scattering", IEEE Phot. Tech. Lett., vol.5, no.1, pp.111-113, May 1989.

[228] N. Edagawa, K. Mochizuki and Y. Iwamoto : "Simultaneous amplification of WDM signals by a highly efficient fiber Raman amplifier pumped by high power semi-conductor lasers", Elec. Lett., vol.23, pp.196-197, 1987.

[229] Weijian Jiang and Peida Ye : "Crosstalk in fiber Raman amplification for WDM systems", J. Lightwave Tech., vol.7, no.9, pp.1407-1411, September 1989.

[230] Ming-Seng Kao and Jingshown Wu : "Signal light amplification by stimulated Raman scattering in an N-channel WDM optical fiber communication system", J. Lightwave Tech., vol.7, no.9, pp.1290-1299, September 1989.

[231] M. S. Kao and J. Wu : "High density WDM systems using post-transmitter fibre Raman amplifier to release Raman crosstalk", Elec. Lett., vol.25, no.21, pp.1457-1459, 12 October 1989.

[232] T. H. Maiman : "Stimulated optical emission in fluorescent solids - Theoretical considerations", Phys. Rev., vol.123, pp.1145-1150, 15 August 1961.

[233] R. J. Mears, L. Reekie, S. B. Poole and D. N. Payne : "Neodynium-doped silica single-mode fibre lasers", Elec. Lett., vol.21, no.17, pp.738-740, 1985.

[234] D. N. Payne, L. Reekie, R. J. Mears and S. B. Poole, I. M. Jauncey, J. T. Lin : "Rare earth doped single mode fiber lasers, amplifiers and devices", Paper FN1, Cleo, San Francisco, June 1986.

[235] a) C. G. Atkins, J. F. Massicott, J. R. Armitage, R. Wyatt, B. Ainslie and S. P. Craig-Ryan : "High gain broad spectral bandwidth erbium doped fibre amplifier pumped near 1.5 µm", Elec. Lett., vol.25 (14), pp.910-911, 1989.

[235] b) JF. Marcerou, S. Artigaud, J. Hervo and H. Février : "Basic comparison between fluoride and silica doped fibre amplifiers in the 1550 nm region", ECOC'92 Proc., vol.1, p.53, 1992.

[236] J. Boggis, A. Lord, M. Violas, D. Heatley and W. A. Stallard : "Broadband high sensitivity erbium amplifier for operation over wide wavelength range", Elec. Lett., vol.26, no.8, pp.532-533, 1990.

[237] W. Sessa, R. Welter, M. W. Maeda, R. E. Wagner, L. Curtis, J. Young, H. Kodera, Y. Koga and R. I. Laming : "Recent progressing multichannel coherent ligthwave systems at Bellcore", Proc. SPIE, vol.1175, pp.241-248, 1989.

[238] A. M. Hill, D. B. Payne, K. J. Blyth, J. W. Arkwrigth, R. Wyatt, J. F. Massicot, R. A. Lobbett, P. Smith and T. G. Hodgkinson : "7203 user WDM broadcast network employing one erbium doped fibre power amplifier", Elec. Lett., vol.26, no.9, pp.605-607, 1990.

[239] M. A. Saifi, Chinlon Lin and W. Way : "Optical fiber amplifiers for broadband optical network applications", FOC LAN 1990, reprinted in Fiber Optics, pp.29-33, Jan.1991.

[240] Chinlon Lin, Winston Way and M. A. Saifi : "Optical fiber amplifiers make broadband fiber networks practical", Laser Focus World, pp.161-171, February 1991.

[241] The editors, Phillips business information, Fiber Optics News, Potomac USA, July 13, 1992.

[242] P. D. D. Kilkelly, P. J. Chidgey and G. Hill : "Experimental demonstration of a three-channel WDM system over 110 km using superluminescent diodes", Elec. Lett., vol.26, no.20, pp.1671-1673, 27 September 1990.

[243] a) Y. Kimura, M. Nakazawa and K. Suzuki : "Ultra-efficient erbium-doped fiber amplifier", Appl. Phys. Lett., vol.57 (25), pp.2635-2637, 17 December 1990.

Bibliography

[243]   b) M. Shimizu, M. Yamada, M. Hariguchi, T. Takeshita, M. Okayasu : "Erbium-doped fibre amplifier with extremely high gain coefficient of 11.0 dB/mW", Elec. Lett., vol.26, no.20, pp.1641-1643, 1990.

[244]   J. L. Zyskind, C. R Randy Giles, J. R. Simpson and D. J. DiGiovanni : AT&T. Tech. J. (USA), vol.71, no.1, pp.53-62, January 1992.

[245]   Paul S. Henry : "High-capacity lightwave local area networks", IEEE Com Magazine, pp.20-26, October 1989.

[246]   Revue Siliconix no.22, Siliconix company 94019 Créteil France, Février 1988.

[247]   A. H. Gnauck, B. L. Kaspers, R. A. Linke, R. W. Dawson, T. L. Koch, T. J. Bridge, E. G. Burkhardt, R. T. Yen, D. P. Wilt, J. C. Campbell, K. Ciemiecki, Nelson and L. G. Cohen : "4 G/bit/s transmission over 103 km of optical fiber using a novel electronic multiplexer/demultiplexer", J. Lightwave Tech., vol. LT3, no.5, pp.1032-1035, October 1985.

[248]   H. M. Rein and R. Reimann : "6 Gbit/s multiplexer and regenerating demultiplexer ICs for optical transmission systems band on a standard bipolar technology", Elec. Lett., vol.22, no.19, pp.988-990, 11 September 1986.

[249]   R. B. Nubling, J. Yu, K. C. Wang, P. M. Asbeck, N. H. Sheng, R. L. Pierson, G. J. Sullivan, M. A. McDonald, A. J. Price and D. M. Chen : "High speed 8:1 multiplexer and 1:8 demultiplexer implemented with AlGaAs/GaAs HBTs", GaAs IC Symposium, CH 2889 - 4/90/0000-0053 - IEEE, 1990.

[250]   Vitesse Camarillo CA commercial documentation, September 1989.

[251]   P. A. Rosher, S. C. Fenning, M. J. Ramsay and F. V. C. Mendis : "Broadband video distribution over passive optical network using subcarrier multiplexing techniques", Elec. Lett., vol.25, no.2, 19 January 1989.

[252]   P. M. Hill : "8 Gb/s subcarrier multiplexed coherent lightwave system", IEEE Phot. Tech. Letters, vol.3, no.8, pp.764-766, August 1991.

[253]   P. A. Davies and Z. Urey : "Subcarrier multiplexing in optical communication networks" Electron. Commun. Eng. J. (UK), vol.4, no.2, pp.65-72, April 1992.

[254]   P. E. Barnsley, G. E. Wickens, H. J. Wickes and D. M. Spirit : "A 4×5 Gb/s transmission system with all optical clock recovery", Photonics Technol. Lett., vol.4, no.1, pp.83-86, 1992.

[255]   G. Farrell, P. Phelan and J. Hegarty : " All optical synchronisation with frequency division self pulsating laser diode ", Elec. Lett., vol.28, no.8, pp.738-739, 9 April 1992.

[256]   D. B. Payne and J. R. Stern : "Technical options for single mode local loops : TDM or WDM ?", ECOC 86, Proceedings, pp.465-468, 1986.

[257] I. J. Goddard and N. Baker : "Multichannel networking and its applications", CH 2538-7/88/0000-0016, IEEE 1988.

[258] S. S. Wagner and H. Kobrinski : "WDM applications in broadband telecommunication", IEEE Communication Magazine, pp.22-30, March 1989.

[259] Rapport de synthèse du groupe optoélectronique de l'observatoire français des techniques avancées, Optoélectronique et réseaux de communications, Masson Ed., 120 bd St Germain, Paris, 1988.

[260] J. A. Bannister and M. Gerla : "Design of the wavelength-division optical network", ICC 90, IEE, vol.3, pp.962-967, April 1990.

[261] F. Blanc, G. Roullet, J. Sourgens et J. Trompette : "Système optique de raccordement d'abonné large bande numérique", Opto 84 Proc., pp.221-223, ESI pub, Paris 1984.

[262] K. Nosu : "Fiber-optic wavelength division multiplexing technology and its applications", Japan annual reviews in electronics computers and telecommunication, vol.5, Y. Suematsu Ed., OHM and North Holland pub, 1983.

[263] W. Rohrbeck : "Four-channel wavelength division multiplexing and its application to the BIGFON project of the German post office", SPIE Proc., vol.468, pp.132-137, Fiber optics'84, 1984.

[264] C. Veyres and J. J. Mauro : "Fiber to the home: Biarritz (1984) ... twelve cities 1988", CH 2538 - 7/88/0000 - 0874, IEEE 1988.

[265] A. Perez : "Solution Velec-CGCT pour les réseaux de vidéocommunication de première génération", Opto 85, ESI Proceedings, pp.10-12, Paris, 21-23 mai 1985.

[266] G. Maes and A. Perez : "The first generation videocommunication network", SPIE Proc., vol.585, p.157, 1985.

[267] Ph. Gacoin, F. Bos, J. Flamand, J. C. Gautherin and J. P. Laude : "High performances WDM devices and applications", Opto 81 Proc., pp.31-36, Masson Ed., Paris,1981

[268] C. Duret : "Centres de distribution dans les réseaux de vidéocommunication", Opto 86, ESI Proceedings, pp.18-22, 1986.

[269] G. Finnie: "International news", Photonic Spectra, p.44, December 1986.

[270] Kohichi Ohta, Yoshio Miyamori, Tsutomu Yoshiga and Minoru Maeda : "Optical subscriber transmission system", Hitachi Review, vol.37, no.3, pp.127-132, 1988.

[271] K. Oguchi, Y. Hakamada and J. Minowa : "Optical design and performance of wavelength division multiplexed optical repeater for fiber optic passive star networks connection", J. Lightwave Tech., vol.LT4, no.6, pp.665-671, June 1986.

[272] Phillips Publishing, Inc Analysis - FTTC 1992.

[273] D. B. Payne and J. R. Stern : "Single mode optical local networks" CH 2190-7/85/0000-1201, IEEE 1985.

[274] M. S. Goodman, H. Kobrinski and K. W. Loh : "Application of wavelength division multiplexing to communication networks architectures", ICC'86, Conference Record, p.931, Toronto, 1986.

[275] R. P. Marsden, J. J. Allen, G. J. Cannell, J. P. Laude and H. Mulder : "A multi-gigabit/s optical business communication system using wavelength and time division multiplexing techniques", ECOC'90 Proceedings, vol.2, pp.779-786, 1990.

[276] J. T. Zubrzycki : "Initial tests on a high-density wavelength division multiplexed network", ECOC'91 Proceedings, pp.345-348, 1991.

[277] D. C. Griffiths, S. C. Thorp, B. R. White, M. J. Hodgson, N. R. Back and J.C.C. Shaw : "A hybrid integrated four-channel wavelength demultiplexing receiver", ECOC'91 Proceedings, pp.305-308, 1991.

[278] N. Shimosaka *et al* : "A photonic wavelength-division and time-division hybrid multiplexed network using tunable wavelength filters for a broadcasting studio application", ECOC'91 Proceedings, pp.545-548, 1991.

[279] S. Wagner, H. Lemberg, H. Kobrinski, L. Sinoot and T. Robe : "A passive phononic loop architecture employing wavelength division multiplexing", CH 2535 - 3/88/0000 - 1569, IEEE 1988.

[280] J. R. Stern, C. E. Hoppit, D. B. Paine, M. H. Reeve and K. Oakley : "Passive optical networks for telephony application and beyond", Elec. Lett., vol.23, no.24, pp.203-206, 19 nov 1988.

[281] C. E. Hoppitt and D. E. Clarke : "The provision of telephony over passive optical network", British Telecom., Technol. J., vol.7, no.2, pp.100-114, April 1989.

[282] P. J. Smith, S. Culverhouse, D. M. Russ and D. W. Faulkner : "A high speed optically amplified TDMA distributive switch network", ECOC'91 Proceedings, pp.89-92, 1991.

[283] D. Newman and J. Stern :"Last link in the loop", Physics world, pp.45-48, September 1991.

[284] A. K. Wood, N. Carr and A. C. Carter : "Design and fabrication of monolithically integrated DFB laser wavelegth duplexer transceivers for TPON/BPON access link", Elec. Lett. (UK), vol.27, no.10, pp.809-810, 19 May 1991.

[285] P. J. Williams, P. M. Charles, R. H. Lord, R. Cloutman, D. Gupta, R. Ogden, F. Randle, S. Thomas, G. M. Foster, T. J. Reid and A. C. Carter : "Demonstrations of optoelectronic integrated circuits in bidirectional TPON/BPON access links", ECOC'91 Proceedings, p.485,1991.

[286] A. M. Hill, D. B. Payne, K. J. Blyth, D. S. Forrester, A. Silvertown, J. W. Arkwright, D. W. Faulkner and J. W. Ballance : "An experimental broadband and telephony passive optical network", IEE Globecom'90, vol.3, pp.1856-1860, 1990.

[287] G. Mogensen, W. Eickhoof, W. Bambach, A. Madani, L. J. Stafg, R. François, J. J. Crosnier, W. Warzanskyj, G. Folke, J. M. Chateau, G. Nilsson and P. Strömgren : "RACE Project R1030 Access - A system study of the broad-band subcriber loop ", J0. Lightwave Tech., vol.7, no.11, pp.1715-1726, November 1989.

[288] J. P. Laude : "Spectromètre à deux Fabry Perot asservis", Thèse, Orsay University, 1966, and Jl Phys, Colloque C2, Sup. au no.3-4, Tome 28, pp.322-325, March 1967.

[289] J. P. Kaminow, P. P. Iannone, J. Stone and L. W. Stulz : "FDM-FSK star network with a tunable optical filter demultiplexer", Elec. Lett., vol.23, no.21,, 8 Oct 1987.

[290] A. E. Willner, J. P. Kaminow, M. Kuznetsov, J. Stone and L. W. Stulz : "FDMA-FSK non coherent star network operated at 600 Mbit/s using two electrode DFB lasers and fibre optical filter demultiplexer", Elec. Lett., vol.25, no.23, pp.1600-1602, 9 November 1989.

[291] H. Toba, K. Oda, K. Nakanishi, N. Shibata, K. Nosu : "Broadband information distribution networks employing optical frequency division multiplexing technologies", CH 2827 - 4/90/0000 - 1512 - IEEE 1990.

[292] D. W. Smith and I. W. Stanley : "The worldwide status of coherent optical fibre transmission systems", ECOC'83, Elsevier Ed., 1983.

[293] R. Gross, P. Hill and R. Olshansky : "Coherent fiber optic subcarrier multiplexed systems", SPIE, vol.1175 - Coherent Lightwave Com., pp.165-167, 1989.

[294] S. Saito, Y. Yamamoto and T. Kimura : "Modulation and demodulation technology for coherent optical fiber transmission FSK heterodyne detection", Review of the Electrical Com. Lab., vol.31, no.3, pp.331-339, 1983.

[295] B. Glance, K. Pollock, C. A. Burrus, B. L. Kasper, G. Eisenstein and L. W. Stulz : "Densely spaced WDM coherent optical star network", Elec. Lett., vol.23, no.17, pp.875-876, 13 August 1987.

[296] B. Glance, O. Scaramucci, T. L. Kock, K. C. Reichmann, L. D. Tzeng, V. Koren and C. A. Burrus : "Densely spaced FDM coherent optical system with random access digitally tuned receiver", Globecom 89, IEEE, pp.343-345, Dallas, 27-30 November 1989.

[297] P. Mols and P. Hooijmans : "Coherent pictures of the future", Physics World, pp.66-70, September 1991.

[298] R. Hüber et al : "Research and development in advanced communications technologies in Europe", Race 91, XIII/142/91, 1991.

[299] C. K. Wong and L. L. Kanters : "Fully engineered transmitter laser units implemented in a coherent multi-channel demonstrator system", ECOC'91 proc., pp.52-59, Paris, September 1991.

[300] Q. Jiang : "Channel selection and identification for coherent optical FDM systems", IEEE Phot. Tech., vol.3, no.8, pp.767-768, August 1991.

Bibliography

[301] K. Smith : "ECOC'82 reprints on coherent fiber optics taps potential for unlimited bandwidth", Electronics, pp.81-88, May 17, 1984.

[302] E. J. Bachus, R. P. Braun, C. Caspar, E. Grossman, H. Foisel, K. Heimes, B. Srebel and F. J. Westphal : "Ten-channel coherent optical fiber transmission", Elec. Lett., vol.22, no.19, pp.1002-1003, 11th September 1986.

[303] D. Capolupo and F. Del Castello : "A flexible integrated multiservice optical coherent network based on TDMA technique", EFOC LAN 90 Proc., pp.97-102, 1990.

[304] A. Fioretti, E. Neri, S. Forcesi, O. J. Koning, A. C. Labrujere, J. P. Bekooij, B. Hillerich, E. Weidel and G. Veith : "An evolutionary configuration for an optical coherent multichannel network", Globecom'90 : IEEE Global Telecom. Conf. and Exhibition. Communications : Connecting the Future (Cat. no.90CH2827-4), San Diego, 1990.

[305] K. Iwashita and S. Norimatzu : "Cross-phase modulation influence on a two-channel optical PSK homodyne transmission system", ECOC'91 proc., paper Th C 10, 1991.

[306] M. C. Brain, M. J. Creaner, R. C. Steele, D. Spirit, N. G. Walker, G. R. Walker, J. Mellis, S. Al Chalabi, W. B. Hale, I. C. C. Sturgess, M. Rutherford and D. Trivett : "Field demonstration of coherent WDM with a fibre amplifier repeater, for transparent optical network applications", CH 2827 - 4/90/0000 - 0768 - IEEE, 1990.

[307] Y. C. Chury, K.T. Pollock, P. J. Fitzgerald, B. Glance, R. W. Tkach and A.R. Chralyvy : "WDM coherent star network with absolute frequency reference", Elec. Lett., vol.24, no.21, pp.1313-1314, 13 October 1988.

[308] N. W. Wood, C. A. Park and B. T. Debney : "Transmitter wavelength stabilisation in coherent multichannel systems", Fibre Optics'90, SPIE, vol.1314, pp.37-44, 1990.

[309] S. Ohshima and M. S. Nakamura : "Lightwave frequency-controlling technologies for FDM switching star network", Photonic Switching II, Proc. Int., Top Meeting Kobe, Springer Verlag, pp.254-257, 12-14 April 1990.

[310] J. P. Laude : "RACE 1036 WTDM", Jobin Yvon report, 1990.

[311] R. Noe and K. Drogemuller : "Accurate and simple optical frequency stabilization for coherent multichannel system", Phot. Tech. Lett., vol.4, no.5, pp.505-506, IEEE, May 1992.

[312] M. Grinsec : "La commutation électronique", CNET-ENST, Eyrolles Ed., Paris, 1980.

[313] A. Oliphant, R. Marsden and J. Zubrzycki : "An optical routing system for tomorrow's television studio centers", SMPTE Journal, pp.660-666, July 1987.

[314] K. Y. Eng : "A photonic knockout switch for high speed packet networks", IEEE/IEICE, Global Telecom Conference pp.47.2.1 – 47.2.5, Tokyo, December 1987.

[315] M. S. Goodman : "Multiwavelength networks and new approaches to packet switching", IEEE Communication Magazine, pp.27-35, October 1989.

[316]   a) Romulus, CNET, Opto no.67, p.14, ESI Paris, Mai-juin 1992.

[316]   b) J. Ph. Guignard, A. Hamel, H. Paciollo, G. Fressy and A. Tromeur : "Improvement of an Ethernet network using multiwavelength techniques", EFOC LAN Proc., pp.346-349, Paris, June 1992.

[317]   H. Kobrinski : "Crossconnection of wavelength division multiplexed high speed channels", Elec. Lett., vol.23, pp.975-977, 27 August 1987.

[318]   S. S. Wagner and H. Kobrinski : "WDM applications in broadband telecommunication networks", IEEE Communications Magazine, pp.22-30, March 1989.

[319]   P. J. Chidgey and G. R. Hill : "Experimental demonstration of wavelength routed optical networks over 52 km of monomode optical fiber", Elec. Lett., vol.25, pp.1451-1452, October 1989

[320]   G. G. Voyevodkin, A. A. Kuznetsov and S. M. Nefedov : "A passive optical switch for a fiber-optic subscriber network with spectral multiplexing", Radioteckhnika, no.7, pp.65-67, 1988.

[321]   H. J. Westlake, P. J. Chidgey, G. R. Hill, P. Granestrand, L. Thylen, G. Grasso and F. Meli : "Reconfigurable wavelength routed optical networks : a field demonstration", ECOC'91 Proc., pp.753-756, Paris, 9-11 Sept 1991.

[322]   G. R. Hill, P. J. Chidgey and J. Davidson : "Wavelength routing for long haul networks", IEEE, CH 2655-9/89/0000-0734, 1989.

[323]   K. Hagishima and Y. Doi : "An optical self-routing switch using multiwavelength", IEEE, CH 2655-9/89/0000-0745, 1989.

[324]   A. S. Acampora : "A multichannel multihop local lightwave network", Globecom'87 Proc., pp.1459-1467, 1987.

[325]   M. G. Hluchyjand, M. J. Karol : "Shuffle Net : An application of generalized perfect shuffles to multihop lightwave networks", J. Lightwave Tech., vol.9, no.10, pp.1386-1397, October 1991.

[326]   E. Arthurs, J. M. Cooper, M. S. Goodman, H. Kobrinski, M. Tur and M. P. Vecchi : "Multiwavelength optical crossconnect for parallel processing computers", Elec.Lett., vol.24, no.2, pp.119-120, 21st January 1988.

[327]   H. Kobrinski, E. Arthurs, R. M. Bulley, J. M. Cooper, E. L. Goldstein, M. S. Goodman, and M. P. Vecchi : "An optoelectronic packet switch utilizing fast wavelength tuning", IEEE, CH 2535 - 3/88/0000-0948, 1988.

[328]   E. Arthurs, M. S. Goodman, H. Kobrinski and M. P. Vecchi : "Hypass : an optoelectronic hybrid packet switching system", J. Selected Area in Commun., vol.6, pp.1500-1510, 1988.

[329]   M. C. Brain and P. Cochrane : "Wavelength-routed optical networks using coherent transmission", IEEE, CH 2538 - 7/88/0000-0026, 1988.

Bibliography 197

[330] S. Suzuki, M. Fujiwara, S. Murata : "Photonic wavelength division and time division hybrid switching networks for large line capacity broadband switching systems", IEEE, CH 2535 -3/88/0000-0933, 1988.

[331] N. Shimosaka, M. Fujiwara, S. Murata, M. Henmi, K. Emma and S. Suzuki : "Photonic wavelength division and time division hybrid switching system utilizing coherent optical detection", IEEE Phot. Tech. Lett., vol.2, no.4, pp.301-303, April 1990.

[332] B. Glance, O. Scaramuci and T. L. Koch : "Random access digitally tuned optical heterodyne receiver", ECOC'89, pp.372-375, Gothenburg, 1989.

[333] Y. Kanayama, N. Goto and Y. Miyazaki : "Integrated optical switches for wavelength division multiplexed communication using colinear acousto-optic interaction", Proc. of the Int. Top Meeting Kobe, Japan, pp.245-248, Springer Verlag, Berlin, April 1990.

[334] A. De Bosio, P. Cinato, B. Costa, A. Daniele and E. Vezzoni : "ATM photonic switching architecture based on frequency switching techniques", Proc. of the Int. Top Meeting Kobe, Japan, pp.300-303, Springer Verlag, Berlin, April 1990.

[335] Y. Takahashi, E. Amada : "A new timing architecture for optical ATM switching systems", Proc. of the Int. Top Meeting Kobe, Japan, pp.304-307, Springer Verlag, Berlin, April 1990.

[336] N. Ogino, M. Fujioka : "Photonic lattice type self routing switching network", Proc. of the Int. Top Meeting Kobe, Japan, pp.312-315, Springer Verlag, Berlin, April 1990.

[337] I. H. White, K. O. Nyairo, C. J. Armistead and P. A. Kirkby : "Crosstalk compensated WDM signal generation using a multichannel grating cavity laser", ECOC'91 Proc., p.689, 1991.

[338] J. Sharony, S. Tiang, T. E. Stern and K. W. Cheung : "Wavelength-rearrangeable and strictly non blocking networks", Elec. Lett., vol.28, no.6, pp.536-537, 12 March 1992.

[339] W. I. Way, D. A. Smith, J. J. Johnson and H. Izadpanah : "A self routing WDM high capacity SONET ring network", IEEE Phot. Tech. Lett., vol.4, no.4, pp.402-405, April 1992.

[340] H. Richter, O. Leminger et R. Ries : "Système de transmission bidirectionnelle à 284/81 Mb/s par fibres optiques pour le projet DOTAN", Opto 84 Proc., pp.219-220, ESI Ed., Paris, 1984.

[341] F. Küffer : "Glasfaserstation 85 zür optischen Übertragung von Sprache und Daten", Bulletin technique PTT, N11/1986, Berne, vol.64, pp.533-538, 1986.

[342] H. Taga, S. Yamamoto, K. Mochizuki and H. Wakabayashi : "2,4 Gb/s 1,55 µm WDM bidirectional optical fiber transmission experiments", Trans. IEICE, vol. E71, no.10, pp.940-942, October 1988.

[343] V. C. Y. So, J. Jiang, D. D. Clegg, P. Valin, F. A. Huszarik and P. J. Vella : "Multiple wavelength bidirectional transmission for subscriber loop applications", Elec. Lett., vol.25, no.1, pp.16-18, 5 January 1989.

[344] S. Celaschi, J. B. Rosolem and J. T. Jesus : "1.3/1.55 micron duplex-diplex optical transmission : the Brasilian technology", CH 2901-7-90-0000-0454 - IEEE 1990.

[345] J. Arnaud : "Ligne de transmission de 6 canaux vidéo sur fibre gradient d'indice à 1300 nm", Opto 85 Proc., pp.30-31, ESI Ed., Paris, 21-23 Mai 1985.

[346] F. V. C. Mendis, T. T. Tjhung and B. Selvan : "A long wavelength WDM system for multichannel TV transmission on optical fibre", IEEE, CH 2175 - 8/85/0000-0021, 1985.

[347] A. Huriau : "Le système Artis, le réseau de transport des programmes audiovisuels", Opto 86 Proc., pp.30-43, ESI Ed.,Paris, 1986.

[348] J. F. Hélard et J. L. Roux : "Transmission optique utilisant le multiplexage en longueur d'onde à 4 DEL", Opto 86 Proc., pp.27-29, ESI Ed.,Paris, 1986.

[349] The editors : "Multiplexage à 4 DEL", Revue Innovation télécom, Centre National d'Etude des Télécommunications, Paris, p.31, hors série 1986.

[350] D. Boisseau, C. Claverie, S. Hergault et D. Thépant : "Transmission simultanée de 4 à 16 programmes vidéo-audio sur fibre monomode", SIRFO Conférence, Paris, October 1990.

[351] M. Bougeot, J. Cellmer, P. Delhange, J. Vieillart et C. Veyres : " Utilisation des fibres optiques pour les liaisons intercentraux, les vidéocommunications et le raccordement d'abonnés à Paris et en Ile de France", Opto 85 Proc., pp.13-16, ESI Ed.,Paris, 1985.

[352] M. Miwa, S. Aoyama, M. Sugita, T. Matsukawa and R. Nakata : "DPCM optical video transmission equipment with wavelength division multiplexing", National Technical Report, vol.33, no.6, pp.40-47, Japon, December 1987.

[353] A. K. Hansson, K. Lindén : "Optical fibre line system for 4x140 Mb/s, a new 565 Mb/s application", Ericsson Review, no.3, pp.102-109, 1987.

[354] The editors, Fiber Optics News, p.3, May 1 1989.

[355] H. Tsushima, S. Sazaki, K. Kuboki, S. Kitajima, R. Takeyari, M. Okai : "1,244 Gb/s 32 channels 121 km transmission experiment using shelf-mounted CPFSK optical heterodyne system" ECOC'91 Proc., p.397, 1991.

[356] K. W. Cheung, S. C. Liew and C. N. Lo : "Experimental demonstration of multwavelength optical network with microwave subcarriers", Elec. Lett., vol.25, no.6, pp.381-383, 16 March 1986.

[357] H. J. Westlake, G. R. Hill, G. E. Wickens and B. P. Cavanagh : "Subcarrier multiplexed transmission using wavelength division multiplexing and optical amplifier", Elec. Lett., vol.25, no.10, pp.632-634, 11 May 1989.

[358] M. A. Santoro, A. Porter, U. Koren, J. Stone and K. Y. Eng : "A 2,4 Gb/s WDM link using a 4-channel integrated tunable laser transmitter package and Fabry Perot filters", ECOC'91 Proc., p.505, Paris, 1991.

[359] P. S. Natarajan, P. S. Venkatesan, C. W. Lundgren and Chinlon Lin : "Semiconductor laser-based multichannel analog video transmission using FDM and WDM over single mode fiber", SPIE, vol.1043, Laser Diode Technology and Applications, pp.260-262, 1989.

[360] G. R. Hill : "Wavelength domain optical network techniques", Proceedings of the IEEE, vol.77, no.1, pp.121-132, January 1989.

[361] J. J. Refi : "LED bandwidth of multimode fibers as a function of laser bandwidth and LED spectral characteristics", J. Lightwave Tech., vol. LT4, no.3, pp.265-271, March 1986.

[362] M. Stern, J. L. Gimlett, L. Curtis, M. K. Cheung, M. B. Romeiser and W. C. Young : "Bidirectional LED transmission on single mode fibre in the 1300 and 1550 nm wavelength regions", Elec. Lett., vol.21, no.20, pp.928-929, 26 September 1985.

[363] D. Mestdagh and I. Van De Voorde : "A 1:N revertive optical protection switching architecture for fiber in the loop systems", ECOC'91 Proc., p.765, Paris, September 1991.

[364] M. Corke, A. Beaudet, P. Dwyer, D. Haynes, R. Kleckowski, R. Moran, D. Werthman and N. Ronan : "Route diversity in fiber networks using optical switching", SPIE, vol.1577, pp.41-51, 1991.

[365] D. H. Hartman : "Digital high speed interconnects : a study of the optical alternative", Opt. Eng., vol.25, no.10, pp.1086-1101, October 1986.

[366] H. D. Hendricks : "Optical backplane interconnection", NASA Tech. Briefs, p.20, February 1991.

[367] D. Z. Tsang : "One gigabit per second free space optical interconnection", Appl. Opt., vol.29, p.2034, 1990.

[368] C. R. Husbands, M. M. Girard and R. Antoszewska : "The application of spectral sliced LED technology to optical interconnects of high speed backplanes", EFOC'91 Proc., pp.386-389, 1991

[369] J. Jannson, F. Lin, B. Moslehi and K. Shirk : "Integrated-optic interconnects and fiber-optic WDM data links based on volume holography", SPIE, vol.1555, pp.159-176, 1991.

[370] I. M. Jauncey, L. Reekie, R. J. Mears, D. N. Payne, C. J. Rowe, D. C. J. Reid, I. Bennion and C. Edge : "Narrow-linewidth fibre laser with integral fibre grating", Elec. Lett., vol.22, no.19, pp.987-988, 11 September 1986.

[371] P. M. Gabla, E. Leclerc and C. Coeurjolly : "Practical implementation of a highly sensitive receiver using an erbium-doped fiber preamplifier", ECOC'91 Proc., p.589, Paris, 1991.

[372] H. Bülow and R. H. Rossberg : "Performances of twisted fused fiber EDFA lump and WDM couplers", ECOC'91 Proc., pp.73-76, Paris, 1991.

[373] M. Zervas and R. Laming : "Erbium doped fibre optical limiting amplifier", ECOC'92 Proc., pp.81-84, Berlin, September 1992.

[374] O. Lumholt, K. Schüsler, A. Grunnet-Jepsen, A. Bjarklev, S. Dahl-Petersen, J. H. Povlsen, T. Rasmussen, K. Rottwitt and C. C. Larsen : "Low noise high gain preamplifier using an isolator within the erbium doped fiber", ECOC'92 Proc., pp.93-96, Berlin, September 1992.

[375] R. Laming, M. Zervas and D. Payne : "54 dB gain quantum noise limited erbium doped fibre amplifier", ECOC'92 Proc., p.89-92, Berlin, September 1992.

[376] H. Schmuck and Th. Pfeiffer : "Fiber pigtailed Fabry Perot filter used as tuning element and for comb generation in an Erbium doped fibre ring laser", ECOC'91 Proc., paper Tu B3-3, Paris, 1991.

[377] Y. Oshishi, T. Kanamori and S. Takahashi : "Concentration quenching in Pr^{3+} doped fluoride fiber amplifier operating at 1.3 µm", ECOC'91 Proc., pp.17-20, Paris, 1991.

[378] a) S. F. Carter, R. Wyatt, D. Szebesta and S. T. Davey : "Quantum efficiency and amplification at 1.3 µm in a Pr^{3+} doped fluorozirconate single mode fibre", ECOC'91 Proc., pp.21-24, Paris, 1991.

[378] b) M. Yamada, M. Shimizu, Y. Ohishi, T. Kanamori, M. Horiguchi and S. Takahashi : "15.1 dB-gain Pr^{3+} doped fluoride fiber amplifier pumped by high power laser diodes modules", ECOC'92 Proc., vol.1, pp.49-52, 1992.

[378] c) Y. Miyajima : "Progress towards a practical 1.3 µm optical fibre amplifier", ECOC'92 Proc., vol.1, pp.687-694, 1992.

[379] W. H. Hatton and N. Nishimura : "New field measurement system for single mode fibre dispersion utilizing wavelength division multiplexing technique", Elec. Lett., vol.21, no.23, pp.1072-1073, 7 November 1985.

[380] W.H. Hatton and N. Nishimura : "New method for measuring the chromatic dispersion of installed single mode fibers utilizing wavelength division multiplexing techniques", IEEE 0733-8724/86/0800-1116, 1986.

[381] M. C. Hutley, R. F. Stevens and D. E. Putland : "Wavelength encoded optical fibre sensors", Sensor Review, pp.64-67, England, April 1985.

[382] K. Fritsch, G. Beheim : "Wavelength division multiplexed digital optical position transducer", Opt. Lett., vol.11, no.1, pp.1-3, January 1986.

[383] B. Jones : "The pig that looks after the railway lines", Sensor Review, vol.6, no.4, p.199-201, IFC Pub Ltd, 0260-2288, 1986.

[384] B. Jones : "Photonic systems meet monitoring and control", C&I, London, pp. 47-51, September 1984.

Bibliography

[385] J. M. Bouchet, J. Meyer and JP Laude : "Procedé et dispositif de conduite et de surveillance d'une installation industrielle par transmission d'in formations et d'ordres par voie optique", European patent 0206901 A1, 10 June 1986.

[386] L. Figueroa, C. Shong, R. W. Huggins, G. E. Miller, A. A. Popoff, C. R. Porter, D. K. Smith and B. Van Deventer : "Fiber optics for military aircraft flight systems", IEEE L.T.S Lightwave Com. Systems, February 1991.

[387] L. Figueroa, C. S. Hong, G. E. Miller, C. Porter and D. K. Smith : "Technology trends : photonics technology for aerospace applications", Photonic Spectra, pp.120-124, July 1991.

[388] T. Hamet, F. Duchateau et D. Colas :"Les systèmes multiplexés par fibre optique adaptés aux applications industrielles", Opto 90 Proc., pp.385-391, Paris, ESI Ed., 1990.

[389] G. Maes : "Réseau Epeire", Opto 90 Proc., pp.179-185, Paris, ESI Ed., 1990.

[390] G. Boisdé, S. Rougeault et J. J. Perez : "Spectrométrie à distance à l'aide de fibres optiques dans le domaine de l'ultraviolet", Opto 86 Proc., pp.71-82, ESI Ed., Paris, 1986.

[391] M. Clément : "Pyromètre multicanalà fibres", Colloque capteurs TEC 88 Grenoble, session no.4, conférence no.3, proceedings edited by association PROCAP, Grenoble, October 1988.

[392] J.P. Laude, D. Lepère, M. Clément and J. Lerner : "Multichannel fiber optic spectrometer for polychromatic pyrometry", OE Lase'89, SPIE reprints, vol.1055, January 1989.

[393] J.P. Laude, D. Birot and M. Lehaître : "An integrated fiber optic spectrometer for oceanographic measurements", Opto 92 Proc., pp.615-616, ESI Ed., Paris, 1992.

[394] A. Ponsot : "Identification des tissus artériels par spectroscopie laser", thesis, Paris XII, 1991 and G. Jarry, private communication, Paris XII University, 1991.

[395] I. Nir and Y. Talmi : "CCD detectors record multiple spectra simultaneously", Laser Focus World, pp.111-120, August 1991.

[396] T. W. Leonard, B. S. Vidula and M. A. Corbin : "The impact of wavelength division multiplexing on fiber optic radar remoting", IEEE CH 1909 - 1/83/0000/0584, 1983.

[397] F. Deborgies and P. Richin : "Fibre's low loss is microwave's gain", Physics World, pp.73-76, September 1991.

[398] E. G. Peck, C. E. Zah and K. W. Cheung : "All-optical image transmission through a single mode fiber", Opt. Lett. (USA), vol.17, no.8, pp.613-615, 15 April 1992.

[399] P. A. Rosher and A. R. Hunwicks : "The analysis of crosstalk in multichannel wavelength division multiplexed optical transmission systems and its impact on multiplexer design", IEEE Journal on Selected Areas in Communications, vol.8, no.6, pp.1108-1114, August 1990.

[400] A. A. Al-Orainy and J. J. O'Reilly : "Error probability bounds and approximations for the influence of crosstalk on wavelength division multiplexed systems", IEE proc., vol.137, Pt. J, Optoelectronics, no.6, pp.379-384, December 1990.

[401] S. Geckeler : "Crosstalk penalties in a bidirectional fiber optic WDM system", IEEE Journal on selected areas in communications, vol.8, no.6, pp.1115-1119, August 1990.

[402] P. A. Humblet and W. M. Hamdy : "Crosstalk analysis and filter optimization of single and double cavity Fabry Perot filter", IEEE Journal on selected areas in communications, vol.8, no.6, pp.1095-1107, August 1990.

[403] A. E. Willner : "SNR analysis of crosstalk and filtering effects in amplified multichannel direct-detection Dense WDM System", IEEE Phot. Tech. Lett. (USA), vol 4, no.2, pp.186-189, February 1992.

[404] Y. Namihira, T. Kawazawa and H. Wakabayashi : "Incident polarization angle and temperature dependance of polarization and spectral response characteristics in optical fiber couplers", Appl. Opt., vol.30, no.9, pp.1062-1069, 20 March 1991.

[405] P. Niay, D. Fredricq, C. Lepers, G. Maes et A. Huriau : "Mise en évidence de la diaphonie Raman dans un système de transmission analogique à multiplexage chromatique sur fibre monomode", Opto, no.47, pp.42-47, ESI Ed., Paris, November 1988.

[406] S. Chi and S. C. Wang : "Maximum bitrate length product in the high density WDM optical fibre communication system", Elec. Lett., vol.26, no.18, pp.1509-1512, 30 August 1990.

[407] J. Hegarty, N. A. Olsson and M. McGlashan-Powell : "Measurement of the Raman crosstalk at 1,5 µm in a wavelength division multiplexed transmission system", Elec. Lett., vol.21, pp.395-397, 1985.

[408] D. Cotter : "Stimulated Brillouin scattering in monomode fibre", J. Opt. Com., vol.4, pp.10-19, 1983.

[409] R. G. Waarts and R. P. Braun : "Crosstalk due to stimulated Brillouin scattering in monomode fibre", Elec. Lett., vol.21, pp.1114-1115, 1985.

[410] A. R. Chraplyvy, D. Marcuse and P. S. Henry : "Carrier induced phase noise in angle modulated optical fibre systems", IEEE J. Lightwave Tech., vol. LT 2, pp.6-10, 1984.

[411] Y. Aoki, K. Tajima and I. Mito: "Input power limits of single mode optical fibers due to stimulated Brillouin scattering in optical communication systems", Journal of Lightwave Com., vol.6, no.5, pp.710-719, May 1988.

[412] R. H. Buckley, E. R. Lyons and G. Goga : "A rugged twenty kilometers fiber optic link for 2 to 18 gigahertz communications", SPIE Proceedings, vol.1371-17, 1990.

[413] R. G. Waarts and R. P. Braun : "System limitations due to four-wave mixing in single mode optical fibres", Elec. Lett., vol.22, no.16, pp.873-875, 31 July 1986.

Bibliography

[414] J. G. Palais, T. Y. Lin and S. Tariq : "Power limitations in fiber-optic frequency division multiplexed systems", Fiber and Integrated Optics, vol.10, pp.75-94, Taylor and Francis Ed., 1991.

[415] A. R. Chraplyvy : "Limitation on lightwave communications imposed by optical fiber nonlinearities", J. Lightwave Tech., vol.8, no.10, pp.1548-1557, October 1990.

[416] D. A. Cleland, A. D. Ellis and C. H. F. Sturrock : "Precise modelling of four wave mixing products over 400 km of step-index fibre", Elec. Lett. (UK), vol.28, no.12, pp.1171-1173, 3 June 1992.

[417] G. R. Walker, D. M. Spirit, P. J. Chidgey, E. G. Bryant and C. R. Batchellor : "Effect of launch power and polarization on four-wave mixing in multichannel coherent optical transmission system", Elec. Lett. (UK), vol.28, no.9, pp.878-879, 23 April 1992.

[418] G. R. Walker, D. M. Spirit, P. J. Chidgey, E. G. Bryant and C. R. Batchellor : "Effect of fibre dispersion", Elec. Lett. (UK), vol.28, no.11, pp.989-999, 1992.

[419] D. A. Cleland, X. Y. Gu, J. D. Cox and A. D. Ellis, Elec. Lett. (UK), vol.28, no.3, pp.307-309, 1992.

[420] J. D. Moores : "Ultra-long distance WDM soliton transmission using inhomogeneously broadened fiber amplifiers", J. Lightwave Tech. (USA), vol.10, no.4, pp.482-487, 1992.

[421] A. Hasegawa : "Amplification and reshaping of optical solitons in a glass fiber -IV : use of stimulated Raman process", Opt. Lett., vol.8, p.650, 1983.

[422] L. F. Mollenauer, J. P. Gordon and M. N. Islam : "Soliton propagation in long fibers with periodically compensated loss", IEEE J. Quant. Elect., vol.QE-22, pp.157-173, 1986.

[423] L. F. Mollenauer, E. Lichtman, G. T. Harvey, M. J. Neubelt and B. M. Nyman, Elec. Lett. (UK), vol.28, no.8, pp.792-794, 9 April 1992.

[424] V. E. Zahkarov and A. B. Shabat : "Exact theory of two dimensional self focusing and one dimensional self modulation of waves in nonlinear media", Zh. Ehsp Teor. Fiz., vol.61, pp.118-134, July 1971 - Sov. Phys. JETP, vol.34, pp.62-69, January 1972.

[425] L. F. Mollenauer, S. G. Evangelides and J. P. Gordon, J. Lightwave Tech. (USA), vol.9, no.3, pp.362-367, March 1991.

[426] M. Ding and K. Kukuchi, IEEE Phot. Tech. Lett. (USA), vol.4, no.5, pp.497-500, May 1992.

[427] Y. Fujii : "Information-maintaining separation of optical pulses employing nonlinearity of silica fiber", Appl. Opt., vol.29, no.6, pp.864-869, 1990.

[428] K. K. Shankaranarayanan, S. D. Elby and K. Y. Lau : "WDMA/subcarrier-FDMA lightwave networks: limitations due to optical beat interference", J. Lightwave Tech. (USA), vol.9, no.7, pp.931-943, July 1991.

[429] E. Wolf : "Invariance of spectrum of light on propagation", Phys. Rev. Lett., vol.56, p.1370, 1986.

[430] E. Wolf and A. Gamliel : "Energy conservation with partially coherent sources which induce spectral changes in emitted radiation", J. of Modern Optics, vol.39, no.5, pp.927.-940, 1992.

[431] R. S. Vodhanel and R. E. Wagner : "Multi-gigabit/sec coherent lightwave systems", CH 2655 - 9189/0000-0444,IEEE, 1989.

[432] E. Modone, G. Parisi and G. Roba : "Ultra low-loss optical fibres for long wavelength systems", CSELT Technical Reports, vol XIII, no.3, June 1985.

[433] P. O. Andersson : "Fiber optics for loop applications, a techno-economical analysis", ISSLS 88, no.11.1, Boston, 1988.

[434] M. De Bortoli and P. Passeri : "Wavelength division multiplexing systems for low cost single mode applications", CSELT Technical Reports, vol XVII, no.4, pp.265-268, August 1989.

[435] M. I. Eiger, H. L. Lemberg, K. W. Lu and S. S. Wagner : "Cost analyses of emerging broadband fiber loop architectures", CH 2655 - 9 - 89/0000-0156, IEEE, 1989.

[436] D. N. Merino : " Accessing technological forecasts for the fiber optic communications markets", IEEE Trans on engineering management, vol.37, no.1, pp.53-55, Feb.1990.

[437] N. A. Olsson and P. A. Andrekson : "Prospect for high bit-rate soliton communication", ECOC'92 Proc., vol.2, pp.746-750, 1992.

[438] N. Henmi, S. Fujita, M. Yamaguchi, M. Shikada and I. Mito : "Consideration on influence of directly modulated DFB LD spectral spread and fiber dispersion in multigigabit-per-second long-span optical fiber transmission systems", J. Lightwave Tech., vol.8, no.6, pp.936-944, June 1990.

[439] A. Elrefaie, M. W. Maeda and Guru : "Impact of laser line width on optical channel spacing requirements for multichannels FSK and ASK systems", IEEE Phot. Tech. Lett., vol.1, no.4, pp.88-90, April 1989.

[440] C. F. Cottingham : "Dispersion-shifted fibre", Lightwave, pp.25-26, Nov. 1992.

[441] Y. Koike : "High-bandwidth, low-loss polymer fibres", ECOC'92 Proc., Mo B1.1, vol.1, pp.679-686, Berlin, 1992.

[442] Young, Kai Chen and Ming Wu, "World newsbreaks", Laser Focus World, p.15, Feb. 1991.

[443] S. V. Chernikov, D. J. Richardson, R. I. Laming, E. M. Dianov and D. N. Payne : "70 Gbit/s fibre based source of fundamental solitons at 1550 nm", Elec. Lett. (UK), vol.8, no.13, pp.1210-1212, June 1992.

[444] G. Sherlock, H. J. Wickes, C. A. Hunter and N. G. Walker : "High speed, high efficiency, tuneable DFB lasers for high density WDM applications", ECOC'92 Proc., pp.225-228, Berlin, 1992.

[445] D. Moss, F. Ye, D. Landheer, P. E. Jessop, J. G. Simmons, H. G. Champion, I. Templeton and F. Chatenoud : "Ridge waveguide quantum-well wavelength division demultiplexing detector with four channels", IEEE Phot. Tech. Lett., vol.4, no.7, pp.756-759, 1992.

[446] P. C. Clemens, R. März, A. Reichelt and H. W. Sneider : "Flat-field spectrograph in SiO_2/Si", IEEE Phot. Tech. Lett., vol.4, no.8, pp.886-887, 1992

[447] T. E. Chapuran, S. S. Wagner, R. C. Menendez, H. E. Tohme and L. L. Wang : "Broadband multichannel WDM transmission with superluminescent diodes and LEDs", IEEE Global Telecom Conf., Globecom'91, vol.1, pp.612-618, Phoenix (USA), December 1991.

[448] G. J. Lampard : "Spectrum slicing of light emitting diodes for distribution in the local loop", A.T.R., vol.26, no.1, pp.56-57, 1992.

[449] N. Ken Reay : "Tunable filters for dense wavelength division multiple access systems", Opt. Eng., vol.31, no.8, pp.1671-1675, August 1992.

[450] M. W. Maeda, J. S. Patel, C. Lin, and R. Spicer : "Novel electrically tuneable filter based on a liquid-crystal Fabry-Perot etalon for high-density WDM systems", ECOC'90 Proc., pp.145-148, 1992.

[451] K. Hirabayashi, H. Tsuda and T. Kurokawa : "Tunable wavelength-selective liquid crystal filters for 600-channel FDM systems", IEEE Phot. Tech. Lett., vol.4, no.6, June 1992.

[452] Y. C. Chung : "Temperature-tuned ZnS etalon filters for WDM systems", IEEE Phot. Tech. Lett., vol.4, no.6, pp.600-602, June 1992.

[453] I. C. Chang : "Collinear beam acousto-optic tuneable filters", Elec. Lett. (UK), vol.28, no.13, pp.1255-1256, June 1992.

[454] M. Kavehrad, F. Khaleghi and G. Bodeep : "An experiment on a CDM subcarrier multiplexed optical-fibre local area network", IEEE Phot. Tech. Lett., vol.4, no.7, pp.793-796, July 1992.

[455] K. Oda, H. Toba and K. Kakanishi : "A tuning technique of 128-channel frequency selection switches for optical FDM information distribution systems", IEEE Globecom'91 Proc., vol.3, pp.1580-1586, Phoenix (USA), Dec. 1991.

[456] T. Numai : "1.5 µm phase-controlled distributed feedback tuneable optical filter", IEEE J. Quant. Elect. (USA), vol.28, no.6, pp.1508-1519, 1992.

[457] S. F. Su, R. Olshansky, D. A. Smith and J. E. Baran :"Flattening of erbium-doped fibre amplifier gain spectrum using an acoustooptic tuneable filter", ECOC'92 Proc., WeP 2.3, vol.1, pp.477-480, 1992.

[458] A. Yu, M. O'Mahony, A. S. Siddiqui : "Gain and noise spectrum properties of gain-shaped erbium doped fibre amplifiers using equalising filters", ECOC'92 Proc., vol.1, pp.481-484, 1992.

[459] A. V. Belov, G. G. Deviatykh, E. M. Dianov, A. N. Guryanov, D. D. Gusovskiy, V. F. Khopin and A. S. Kurkov : "Sm^{3+}-doped fibre application to the spectral filtration in the range of 1530-1570 nm", ECOC'92 Proc., vol.1, pp.485-488, 1992.

[460] Y. H. Lee, J. Wu, M. S. Kao and H. W. Tsao : "Performance analysis of wavelength division and subcarrier-multiplexing (WDM-SCM) transmission using Brillouin amplification", IEE Proceedings J., vol.139, no.4, pp.272-279, Aug. 1992.

[461] P. T. Poggiolini and L. G. Kazovsky : "Starnet : an integrated services broadband optical network with physical star topology", Adv. Fiber Comm. Tech., L. G. Kazovsky (ed.), Proc. SPIE, vol.1579, pp.14-29, 1991.

[462] L. G. Kazovsky, C. Barry, M. Hickey, C. A. Noronha and P. Poggiolini : "WDM local area networks", IEEE LTS, vol.3, pp.8-15, May 1992.

[463] H. Toba and K. Nosu : "Optical frequency division multiplexing systems", Trans. IEICE, vol. E75-B, no.4, pp.243-255, April 1992.

[464] J. Sharony, K. W. Cheug and T. E. Stern : "The wavelength dilatation concept-implementation and system considerations", ICC'92 Proc., paper 330.3, Chicago, June 1992.

[465] J. Sharony, K. W. Cheung and T. E. Stern : "Wavelength dilated switches (WDS), a new class of high density, suppressed crosstalk, dynamic wavelength-routing crossconnects", IEEE Phot. Tech. Lett., vol.4, no.8, pp.933-935, August 1992.

[466] R. Rokugawa, N. Fujimoto, T. Horimatsu, T. Nakagami and H. Nobuhara : "Wavelength conversion laser diodes application to wavelength division photonic cross-connect node with multistage configuration", Trans. IEICE (Japan), vol. E75-B, no.4, pp.267-274, April 1992.

[467] M. R. Phillips, A. H. Gnauck, T. E. Darcie, N. J. Erigo, G. E. Bodeep and E. A. Pitman : "112-channel split-band WDM lightwave CATV system", IEEE Phot. Tech. Lett., vol.4, no.7, July 1992.

[468] J. M. Lerner and J. P. Laude : "New vistas for diffraction gratings", Electro-Optics, pp.77-82, May 1983.

[469] A. P. Goutzoulis, D. K. Davies and J. M. Zomp : "Hybrid electronic fiber optic wavelength-multiplexed system for true time delay steering of phased array antennas", Opt. Eng., vol.31, no.11, pp.2312-2322, November 1992.

[470] D. K. Davies and A. P. Goutzoulis : "Wavelength-multiplexed analog fiber optic link for signal transmission", Opt. Eng., vol.31, no.11, pp.2323-2329, November 1992.

[471] R. Comuzzi, C. De Angelis and G. Gianello : "Improved analysis of the effects of stimulated Raman scattering in a multi-channel WDM communication system", Eur. Trans. Telecom. Relat. Technol. (Italy), vol.3, no.3, pp.295-298, May-June 1992.

[472] J. C. Newell : "The use of an optical amplifier for extending the transmission distance in a WDM network", SG1-Optical Networks Workshop Proc., Brussels (RACE), January 13-14, 1993.

[473] T. L. Koch and J. E. Bowers : "Nature of wavelength chirping in directly modulated semiconductor lasers", Elec. Lett., vol.20, pp.1038-1040, December 1984.

[474] J. Buus : "Dynamic line broadening of semiconductor lasers modulated at high frequencies", Elec. Lett., vol.21, pp.129-131, February 1985.

[475] F. Koyama and Y. Suematsu : "Analysis of dynamic spectral width of Dynamic-Single-Mode (DSM) lasers and related transmission bandwidth of single-mode fibres", IEEE, J. Quant. Elect., vol. QE 21, pp.292-297, April 1985.

[476] M. Osinski and J. Buus : "Linewidth broadening factor in semiconductor lasers - an overview", IEEE, J. Quant. Elect., vol. QE 23, no.1, pp.14-25, January 1987.

Index

Aberrations, 59,63,66
Access RACE, 136-138
Acousto-optical, 112-114
Amplifiers (optical), 119-126
Aperture, 17-18
ASK, 139-140

Bigfon, 149
Brillouin, 120,164-165

CCD, 111,157-158
Chemical etching, 65,69
Coherence of source,170
Coherent detection, 139-144,147
Coherent interaction
Coherent λ switch, 148
Coupling, 68-72
   interaction length, 69-70
   spectral selectivity, 22,28,38,66-70
Crosstalk, 25
Cutoff, 7

Demultiplexer, 19,77
Detector, 26,77
Diffraction orders, 38-41
Diode array, 117
DOTAN, 149

Eikonale, 16
Emittor, 21-24,146
  light emitting diodes, 22-23,74
  laser, 23-24,75-76
  thermal shift, 23,28,67
  tuneable, 115,147,169

Fabry-Perot, 111-112,119-120
Filters, 27-36
  LWPF, 27-28
  SWPF, 27-29
  Fabry-Perot, 28,30
FSK, 141-143,148,151
Fusion, 69-71

GaAs, 21,74,77,87
Grating, 37,106-111
  blaze, 39-44,47
  dispersion, 38
  efficiency, 39-47
  principle, 37
  resolution, 38

InP, 21,78,87

Lambdanet, 134
Lasers, 23-24,75-76,119-125,153,167
Lithium niobate, 78,113
Lithium tantalate, 78
Losses, 20-25

Near end crosstalk, 26
Neural network, 116

Optical fibres
  fluoride glass, 21,122
  plastic, 21,155
  silica, 12,20,22,122,125
  single-mode, 3
  step-index, 4
  graded-index, 9-10,16
  DIC, 12
  spectral transmission, 22
Optical frequency multiplexing, 1,138-139

Polarization, 113-115,164
PON/TPON/BPON, 135
PPL, 135
Propagation, 3-18
  dispersion, 11,155
  modes, 4
  Gaussian approximation, 7
  geometrical approximation, 15-18
PSK, 139,143-144

Index

Radars, 161
Raman, 121,164
Romulus, 147

Shuffle net, 147
Silicon, 84-87
Soliton, 13-14,169-170
Star-track, 148
Sub-carrier, 140

Telecommunication networks
　different types, 145-148
　bit rates, 127
Thermal drift, 23,28,67,93,111,120
Time division multiplexing, 128-129
Transmission function, 28-30,55,71

Uncertainty, 165-166

Wavelength division multiplexing
　principle (general), 22
　with diffraction grating (principle), 37
　on car, 155-156
　in process control, 155
　in spectroscopy, 157
　historic, 1, 172-202

WTDM RACE, 133-134,143

Normandie Roto Impression s.a.
61250 Lonrai
N° d'imprimeur : I3-1564
Dépôt légal : août 1993

MASSON Éditeur
120, bd St-Germain
75280 Paris Cedex 06
Dépôt légal : septembre 1993